by Roger Huntington

ACKNOWLEDGMENTS:

A special thanks to the following people for their
contributions to this book:

To Vince Piggins of Chevrolet Product Promotion for checking the Chevrolet material, and for adding background color on Hudson;

To Chuck Condron for wheeling out his post-World War I Fronty Ford and Rajo-head T;

To Steve Christ for setting the record straight on big-block Fords;

To Duane DeButts and Chrysler's Direct Connection for early-Chrysler Hemi information;

To Caroline Panzeter of Hurst Performance for dipping into the Shifty Doctor's files;

To Bob Stevens of Cars & Parts magazine for never turning a deaf ear on numerous inquiries;

To Walt Haessner for unearthing early stock-car-racing photos;

To the many individuals, car clubs and public-relations people who supplied numerous photos and drawings;

And to Cathy Bevers, Ed Francis, Ed Giola, Gary Hanson, Dick Leddy, Larry Ross, and Tom Whitt for stilling their trusty supercars long enough to achieve photographic immortality.

Co-Publishers: Bill & Helen Fisher
Executive Editor: Rick Bailey
Editorial Director: Tom Monroe, P.E., SAE
Editor: Ron Sessions, ASAE
Art Director: Don Burton
Book Design: Paul Fitzgerald
Photos: Roger Huntington; others as noted

Front-Cover Photos: Cobra 427 courtesy of Shelby-American Club; 4-4-2 convertible courtesy of Oldsmobile; Plymouth Superbird courtesy of Classic '60s magazine; and Camaro ZL-1 by Bill Porterfield.
Back-Cover Photos: 1937 Cord by Edward Lane; Shelby G.T. 350 Mustang by Rick Kopec; Chrysler 426 Hemi courtesy of Chrysler Direct Connection; and 1964 GTO convertible by Ron Sessions.

Published by HPBooks, P.O. Box 5367, Tucson, AZ 85703
602/888-2150
ISBN 0-89586-221-2
Library of Congress Catalog number: 82-84044

Contents

Preface

Supercar Superbird. You couldn't miss this Hemi-powered, winged monster on the streets or on the NASCAR high-banked ovals. Photo courtesy of Classic '60s magazine.

WHAT IS A SUPERCAR?

A *supercar* could be loosely defined as a standard passenger car that's been hopped up by the manufacturer with special equipment. The equipment improves the supercar's performance and handling over the standard car.

As defined, such a supercar wouldn't necessarily be very exciting. But the way this idea evolved in the American auto industry in the 1960s was a phenomenon with no parallel in world automotive history.

By a most-bizarre combination of circumstances, four contemporary factors collided to bring it all about: (1) Gasoline was priced extremely low in relation to wages, (2) An automobile was considered a status symbol, (3) The 18—30-age group was becoming a sizable market segment with strong buying power, and (4) The most-popular and fastest-growing classes of auto racing were using basically stock bodies and engines.

These factors combined to usher in a decade of intense factory development of youth-oriented, high-performance cars. Based on standard-production models, American supercars cars eventually delivered speed and acceleration figures equal to the world's most-exotic hand-built cars. Fuel-economy figures were equally outrageous—as low as 6—8 mpg in normal driving. Sold off the showroom floor, these cars cost only a few-hundred dollars more than corresponding bread-and-butter family models. A high-school graduate, only a few weeks on a new job, could afford one.

The natural result was that everyday street driving became a sport. There was a potential race at every traffic light. Status was measured by horsepower. Evenings were spent cruising the town's main street, looking for automotive action. You could cruise for hours on a couple of dollars worth of gas. Your car was your castle. And brute performance was king.

Times are much different now. The swift rise of gasoline prices in recent years has seen to that. Cars can still be fun—but today the manufacturers are also promoting handling and styling, rather than just straight-line speed and acceleration. Even the most-optimistic car enthusiast doesn't expect the wild days of the 1960s to return. We may never see those 0—60-mph figures on factory models again. It's a wonder the supercar ever happened!

Driver's-eye view of an early Mercer Raceabout, circa 1912. You were right out in the open then. Even 50 or 60 mph was a thrill. Photo by David Gooley.

1

Fun-Car Pioneers

Duesenbergs to hot rods

WHEN MEN WERE MEN

It could be said that all of the earliest American automobiles were sports cars. One drove more for fun than for practical transportation in the early days. If you wanted to be sure of getting there, you took a horse. A car was strictly for putting around on a nice day. It was smoky, noisy, jerky and unreliable. The word *performance* hadn't yet been thought of in relation to an automobile. Keeping a car moving under its own power was a real challenge. Motoring was a he-man's game in the early days of the 20th century.

But there was no way this situation could last very long. Even before 1910, there were a few cars in the showrooms that would go a little bit faster and accelerate a little bit quicker than the average car. *Performance* began creeping into the salesman's vocabulary.

At first, performance meant nothing more than the car's ability to climb a steep hill—and there were a lot of them on early roads. A good grade would frequently stop an average car cold. Sometimes the driver could turn around and back up the hill, if reverse-gear ratio was lower than first gear—*low-forward.* But how humiliating! Especially when the next guy, in a more-expensive car with a bigger engine, could make it in low.

Above: Mercer Raceabout of 1912 was perhaps America's first true sports car. It had a relatively sophisticated suspension system, a four-speed transmission, and would go 70 mph with 60 HP. Only a few remain today. Photo by Ted Koopman.

Mercer Raceabout—Undoubtedly, America's first all-out sports car was the 1911 Mercer Raceabout. It was designed strictly for performance and fun driving. The Raceabout was little more than two bucket seats perched on a channel frame, with skimpy fenders, and a huge oval gas tank behind the seats. Its front engine was covered by a hood with leather-belt hold-downs. Frame kickups at each end reduced ride height. The car rode on big 30-in. wood-spoke wheels.

The four-cylinder, T-head engine displaced 300 cubic inches. It developed 60 HP at 2000 rpm. The rev limit was a lazy 2500 rpm. An advanced feature was a close-ratio, four-speed transmission, crash-shifted by a 2-ft-long lever outside the frame. Close-ratio gearing was said to help acceleration.

The manufacturer also *guaranteed* all Raceabouts to cover the flying mile in 51 seconds, at 70 mph. The car might have been faster with the right final-drive ratio, but the designers purposely used a loafing 2.50:1 axle ratio to keep maximum high-gear revs well under 2000 rpm. They knew Raceabout owners would be driving flat-out most of the time!

Stutz Bearcat—The 1914 Stutz Bearcat was a car similar in concept and design to the Mercer Raceabout. It had a slightly larger engine, a little more power and went a little faster. But the idea was the same: a fun-to-drive, two-seat personal sports car with superior performance. Even before World War I, there was a substantial demand for cars of this type.

The Lozier Company supplied elaborate bodies for their early sports cars. This is the 1913 model, with outside "mother-in-law" seat. These were very expensive cars. Photo by Ted Koopman.

Chuck Condron's '23 Ford Model T with single-overhead-cam Rajo head. This modern-day example is indicative of the hopped-up jobs of the '20s. Photo by Tom Monroe.

Other companies with sporty offerings included National, Lozier, Moon, Marion and Simplex. By World War I you could buy a car off the showroom floor that would go 80 mph—on a decent road—and accelerate from a standing start to 60 mph in less than 25 seconds.

There was only one flaw: These early sports cars cost too much for the average man. They weren't just men's cars—they were *rich-men's* cars. The man in the street could only stand and drool. And he wanted to go fast, too.

THE FIRST HOT RODS

Car enthusiasts who think the *hot-rod sport*—modifying a bread-and-butter production car for better performance—started after World War II are wrong. Actually, it goes way back to Model T Fords of the World War I period.

If you think about it, the development of hot rods was inevitable. Henry Ford was grinding out Model Ts by the millions at prices nearly everyone could afford. Parts were dirt-cheap and available everywhere. And the junkyards were full of serviceable Ts and spare parts you could buy for a few dollars. When you combined this with strong grass-roots demand for fun driving and superior performance, the custom Model T speedster was an inevitable outcome.

One other essential ingredient was a healthy, aggressive aftermarket to supply the hundreds of special parts needed for the transformation. Without these parts, those unexcit-

Typical home-built Ford Model T speedster of the World War I period. There was a huge special-parts aftermarket at that time, enabling the enthusiast to build an 80-mph hot rod for a low price. Photo by David Gooley.

1923 T with 16-valve double-overhead-cam Roof cylinder head. Counterweighted crankshaft and pressurized lubrication system helped the T handle increased power. Photo by Frank Hoiles.

Another view of '23 T with Roof DOHC head shows the Winfield SRBD 1-barrel downdraft carburetor, Packard oil pump and Dodge water pump. Photo by Frank Hoiles.

ing Ford production cars would not have become the fire-breathing little speedsters that revolutionized the sport of fun motoring between 1915 and 1925. Although they were called *specials* then, this was the true beginning of the hot-rod sport.

In fact, the aftermarket in special parts to customize Model Ts during this period was far more extensive than the classic hot-rod market in Ford flathead-V8 stuff that developed after World War II. You could buy virtually anything for a Model T in those early days: all kinds of engine hop-up parts, pump-operated cooling systems, suspension components, two-speed rear axles, special wheels, brakes and transmission conversions. Add to this a fantastic choice of special speedster bodies, seats, steering wheels, instruments and windshields. The list goes on and on.

You could end up with a homemade car that contained practically no original Ford parts. I'm looking at a magazine article of some years back that describes one of these custom Ford speedsters of around 1920. Some of the equipment mentioned: a Langdon aluminum body, Ruckstell two-speed rear axle, Buffalo 20-in. wire wheels, Franklin steering, Rajo rocker-arm cylinder heads, Winfield carburetors, Union forged-steel crankshaft, RayDay aluminum pistons and coil-type ignition. These special engine goodies could easily raise output from the T's stock 20 HP at 1700 rpm to 50—60 HP at 3000 rpm. With a light aluminum body, total car weight was well under 1500 lb.

Result: Road performance equal to the expensive production sports cars of the day—at a price the man in the street could afford. Yes, it meant a lot of time and dirty work in your backyard garage. But that's hot rodding. It was as true then as it is today.

Incidentally, this sport of modifying a low-priced production car for better performance didn't die with the Model T. It continued right on with the Model A Ford in the late '20s and early '30s. After World War II, it formed the backbone of the modern hot-rodding movement with the flathead V8 engine. I'll discuss this in more detail later.

One of the more-popular specials of the '20s was the Frontenac, or Fronty Ford. These were made by the Chevrolet Brothers after they sold their auto company to General Motors. Photo by Tom Monroe.

Details of Chuck Condron's '25 T Fronty engine. Overhead-valve head boosted compression to 8.5:1—about double that of a stock flathead Model T. Boring and stroking increased displacement from 176 to 194 cubes. Crank was drilled for pressure lube to mains and rod big ends. Rockers had pressure lube too. Photo by Tom Monroe.

Duesenberg two-seat speedsters were the epitome of sophisticated road performance in the '30s. Supercharged models could approach 120 mph, with brisk acceleration—for around $15,000.

Duesenberg Speedsters of the '30s had a huge straight-8 engine with dual overhead cams and four valves per cylinder. This one had the optional supercharger, which bumped output to more than 250 HP at 4000 rpm.

MORE FUN CARS FOR THE RICH

Despite the obvious popularity of home-built Model T hot rods in the World War I period, American car manufacturers failed to take the hint and develop low-priced sports cars for a broad market segment. Through the 1920s and into the late '30s, America's top-performing cars continued to be relatively high-priced models for the wealthy.

Cunningham V8—But there were some great cars in this high-priced group. One of the most respected in the early '20s was the Cunningham-V8 boattail speedster. Its huge 442-cubic-inch-displacement (CID) L-head engine developed a modest 90—100 HP at 2400 rpm. But with big wheels and a loafing gear ratio, it would fly down the road. A stripped stock model averaged 91.1 mph for five laps on the Sheepshead Bay board speedway, under American Automobile Association (AAA) supervision. The car might have reached 85 mph with full equipment, though acceleration was a little sluggish because of its weight—in excess of 4000 lb. Incidentally, its compression ratio was only 4.3:1, so it would burn just about any fuel that would pour!

Duesenberg Model J—Certainly America's most-famous high-performance luxury car was the Duesenberg Model J, introduced in 1928. Fred Duesenberg had produced an advanced overhead-cam, straight-8 luxury car in the early '20s that had pretty decent performance. But he could never have tackled an exotic, cost-no-object project like the Model J without the big-money backing of a major corporation.

This chance came when E.L. Cord bought out Dusenberg's little company in 1926. Cord assigned Duesenberg the task of designing the finest motor car that could be built with current technology. Projected prices were to range between $12,000 and $20,000. To put that sum of money into perspective, consider this: Henry Ford was paying his assembly-line workers five dollars a day— considered a *good salary!* At that rate, it would take a Ford worker 15—30 years to buy a "Duesie."

Duesenberg engines were painted a distinctive green color, with polished aluminum covers and accessories. As impressive under the hood as outside!

You've probably read the exciting specs many times. The engine followed Fred's design themes for his Indianapolis race engines: Straight-8 with dual overhead camshafts, 4 valves per cylinder, aluminum pistons, tubular connecting rods, with high-lift, long-duration cams and large-bore carburetion. With a displacement of 420 cubic inches, Fred advertised the J engine at 265 HP at 4200 rpm. Actual test curves on production engines showed a true output closer to 200 HP at 3600. But there's no ques-

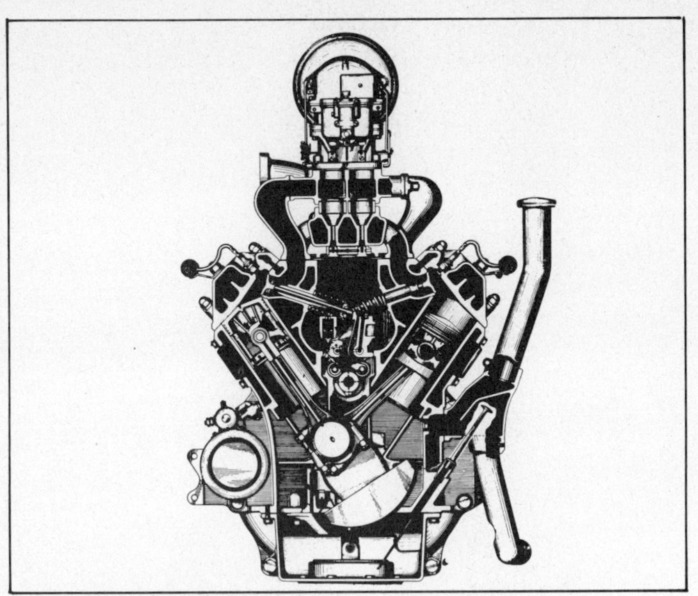

Many luxury cars of the '30s used 12- and 16-cylinder engines with huge displacements. The big engines delivered exceptional performance with no fuss and noise. This Packard V12 had 463 cubic inches. Power was 175 HP at 3200 rpm.

1932 Auburn V12 boattail speedster had a guaranteed top speed of more than 100 mph. Manufactured by the Cord Corp., it offered luxury and image at a medium price.

tion this was the most-powerful engine in U.S. automotive history until the 1950s. Later, in fact, some J engines were fitted with an optional gear-driven centrifugal supercharger. This pumped 5-psi boost pressure at 4000 rpm, adding another 50 or 60 HP.

It's hard to separate fact from fiction on the Model J's road performance. Most auto enthusiasts have never seen a Model J, yet they've been talked about for 50 years. Myths abound. But it's probably safe to say the short-wheelbase, supercharged speedsters would approach 120 mph with the right gear ratio. The unblown, lightweight speedsters were good for a little over 110 mph. Acceleration wasn't blinding. Even the lightest models weighed close to 5000 lb. But one reliable source quoted a 0—60-mph time of 12 seconds for a blown lightweight. Seems possible.

America's most-famous speed-record man, Ab Jenkins, used a specially modified, supercharged Duesenberg to set a whole flock of world records on the Bonneville Salt Flats in 1935. The engine was hopped up to a reported 350 HP at 5000 rpm with dual carburetors, higher compression and reground cams. With a narrow streamlined body fitted to the stock Model J chassis, Jenkins hit speeds up to 152 mph for short distances. But his 24-hour record was only 135 mph. The engine couldn't exceed 4500 rpm for any length of time without destroying its bearings.

Even though the Model J Duesenberg eclipsed the performance of all other American cars in its time, its $12,000—20,000 price was so high that only a handful of very wealthy people could ever hope to own one. In fact, only about 470 of all body styles were assembled in the eight years of production. And there were only two of the short-wheelbase, blown speedsters—custom-made for Clark Gable and Gary Cooper.

In this atmosphere, there was plenty of room for a generation of less-exotic luxury speedsters costing one-third to one-half the price of a Duesenberg. A formula emerged in the '30s for developing such cars: a lightweight-speedster body fitted to a luxury chassis, powered by a 12- or 16-cylinder engine. A number of companies made these chassis: Packard, Cadillac, Auburn, Lincoln, Pierce-Arrow and Marmon. The huge multicylinder engines displaced as much as 500 cubic inches. They could deliver 150—200 HP with very mild camming and carburetion. With a lightweight streamlined body, it was no trick to hit 100 mph and turn 0—60 times of 18—20 seconds. These were still very expensive cars by current standards, but more affordable than a Duesenberg.

Performance remained a rich-man's game.

THE ROOTS OF STOCK-CAR RACING

Many car fans think big-time stock-car racing began with southern-style NASCAR bashes after World War II. Wrong! There was quite a flurry of interest in stock-car racing under AAA supervision in the late 1920s. AAA officials were getting concerned about the high cost of running the tiny 91-CID (1.5-liter) supercharged Millers and Duesenbergs in Championship racing. They wanted to promote some form of grass-roots racing to utilize the beautiful board speedways that had sprung up around the country. The idea of racing factory-stock cars was born.

The chief stock-car adversaries turned out to be Stutz, Studebaker and Auburn, all fielding their popular sport roadsters with straight-8 engines of around 115 HP. Rules permitted stripping off windshields, fenders and running boards, but engines were supposed to use factory equipment only. Top speeds were said to be within 10 mph of the standard showroom model.

The racing turned out to be fast and exciting. The first 75-mile sprint race on the Atlantic City track was held in

Cord Corp. stylist Gordon Beuhrig did wonderful things in the 1930s. No other U.S. company ever approached these flowing lines with the boattail theme. This is a 1934 Auburn. Photo by David Gooley.

1927. It was won by a Stutz at an average speed of 86.2 mph. An Auburn was 2nd, one car-length back, and a Paige roadster was one car-length behind the Auburn. Some wild racing was seen that season, with honors divided between Stutz, Auburn and Studebaker. Auburn and Studebaker also used their team cars to set certified U.S. stock-car speed records on the board tracks. In tests at Daytona Beach, the stripped track cars could hit 104—105 mph.

Late in the '27 season, Stutz brought out a new Blackhawk Speedster model with a more-streamlined body. The Blackhawk also enjoyed a 20-HP increase in power over other Stutz models through changes in carburetion, camming and higher compression. Track speeds jumped 8—10 mph overnight—and Stutz began winning everything.

The other competitors quickly withdrew, and America's short-lived honeymoon with stock-car racing flickered out. But it left a taste for a new type of auto racing with performance fans. And incidentally, Stutz sent a Blackhawk to France in June, 1928 to run in the famous LeMans 24-hour race. To everyone's astonishment, driver Brisson was leading with only 90 minutes to go—then the car's three-speed transmission failed. A Bentley passed him, but he was able to finish 2nd by holding the car in high gear.

No, the Blackhawk wasn't a low-priced, or even medium-priced, sports car. It cost $4900—eight or ten times what you'd pay for a new Ford or Chevy. The common man could only dream.

MEDIUM-PRICED SPORTS CARS

It's a little ironic that the man who gave Fred Duesenberg the green light to design the most-exotic luxury car in the world was also the man who revolutionized the medium-priced, high-performance sports market in the mid-'30s. E.L. Cord, always an astute businessman, was looking for an unfilled niche in the market—one that would appeal to a broad buyer group. The plan was to build

Supercharged Auburn Speedsters of the '35—'36 period gave 100+ mph performance for around $2200. One of the best performance bargains of all time.

Functional sidepipes on a '35 Auburn Speedster.

Blower on a '35 Auburn Speedster. Engine was rated 150 HP at 4000 rpm.

an exciting car based on his front-wheel-drive L-29 Cord of '29—'30. That car had beautiful styling, but it had engineering problems—and it cost too much in a depression economy. The L-29 was a sales failure, but Cord was determined to learn from his mistakes.

Cord's new idea was a series of cars in the $2000—$4000 price range that would have absolutely flamboyant styling. To back up the image the styling promised, the cars would have radical engineering features—but with very careful design and development. Also a must was road performance and handling superior to practically anything manufactured in that day. Cord insisted that styling, both inside and out, not only be exciting and different, but strongly complement the sports *image* of the cars. Cord didn't know how it could be done—but he figured there must be a way to make lines, shapes and gadgets tell a story.

He might never have succeeded in his project if he hadn't stumbled across Gordon Beuhrig, a young Detroit designer. This guy was a genius in projecting an image with lines. Cord gave Beuhrig a free hand. The magnificent Auburn Speedsters of the '35—'36 period and the '36—'37 front-drive Cords were the result. No car lover can help but be excited by the flowing lines of these cars: sweeping clamshell fenders, bold grilles, long hoods, the tapered boattails on the Speedsters, and exotic outside exhaust pipes on the supercharged models. Even the Cord's instrument panel was a work of art. Done in an aircraft-like style, it continued the sporty exterior lines to the inside of the car, where driver and passengers could *feel* the kind of car they were in.

Cord's insistence on advanced engineering assured performance to complement the looks. Weight was kept below 4000 lb, and an optional centrifugal supercharger was offered to give buyers the ultimate in speed and accel-

Ab Jenkins drove a supercharged Auburn Speedster on the Bonneville Flats in 1935 to set many long-distance stock-car records. He averaged 102.90 mph for 12 hours, hit speeds as high as 105 mph for short distances. Photo courtesy of Gulf Oil.

Front-wheel-drive Cords of the '36—'37 period were years ahead of their time in styling and engineering. Photo courtesy of Edward Lane.

Machine-turned '36—'37 Cord dashboard was just as distinctive and sporty as the rest of the car. It carried car's image inside to the driver. Beautiful styling job.

Famous Cord front-wheel drive of '36—'37 had a sophisticated independent front suspension. Transverse leaf spring and trailing arms are shown here. Handling and ride quality were outstanding for that day.

eration. In the Auburn Speedster, the blown, 280-CID straight-8 engine developed 150 HP at 4000 rpm. At AAA stock-car speed runs at the Bonneville Salt Flats, the supercharged straight-8 Speedster recorded a top speed of 105 mph.

Cord also developed a supercharged 289-CID V8 engine. The blown V8 was rated 175 HP at 4200 rpm. The engine pushed the car to a top speed of 108.3 mph in AAA record runs—and this was with fenders! The cars were *strictly stock*. A British car magazine tested a supercharged-V8 Cord for acceleration and got a 0—60-mph time of 13.2 seconds. Standing-start acceleration was hampered by an electro-vacuum gear-shifting system that took its sweet time. But the Cord had excellent passing acceleration on the highway.

It's unfortunate that the Cord Corporation failed financially before this new type of upper-medium-priced fun car could be fully developed. It was a type of car that had always been popular in Europe and Britain. I consider these Cord products to be the first American supercars.

PERFORMANCE FOR THE MASSES

The Ford V8—Auto historians generally agree that Henry Ford's Model T put America on wheels. They also agree that this car was perhaps the most-significant automotive development of the century. What is not quite so obvious

Ford V8 roadsters were the budget pocket rockets of the mid-'30s. Snappy acceleration of this '34 model could embarrass many high-priced sports cars. Photo by Don Randall.

Ford's first V8 model in 1932 revolutionized the performance image of low-priced utility cars. With weight of only 2500 lb, it could out-accelerate many high-priced cars—at least up to 40 or 50 mph.

is that another milestone Ford model had almost as much impact in the performance field—the Model B. The 1932 Model B was the first low-priced car available with a V8 engine.

It's true that Ford's main thrust behind the new V8 was to try to recapture first place from Chevrolet in the sales race. The swift rise of GM in the '20s was a bitter pill for the old man. He was ready to try radical means to battle back. With the V8 engine, he could use "peppy performance" as a major sales theme. The whole car was designed with that idea.

And it was peppy performance. With 65 HP in a car weighing 2500 lb, those first Ford V8s were exciting. Top speed was 80 mph, with a 0—60 time around 18 seconds. Engine output was subsequently improved to 75 HP and then 85 HP, still with car weights under 3000 lb. Everyday passenger-car performance took on new meaning to the average man.

During this period, stoplight street racing began to catch on. The Ford V8 would really scoot from a standing start to 30 or 40 mph. Hardly anything else on the street could touch a Ford in those days. True, the heavier Packards and Duesenbergs could win out on the top end—but those $500 Fords could give them fits for that first block!

Henry's V8 didn't succeed in ousting Chevrolet in the sales race. But it was certainly a success in the performance race.

One example of the sudden superiority of Ford-V8 performance could be seen in the Elgin, Illinois stock-car road races of 1933. The first *seven* places were won by V8 roadsters—stripped of fenders and windshields, but otherwise stock. Winner Fred Frame averaged 80.22 mph for 203 miles. On the main straightaway, slightly downhill, he consistently hit 4900 rpm—100 mph—and then double-

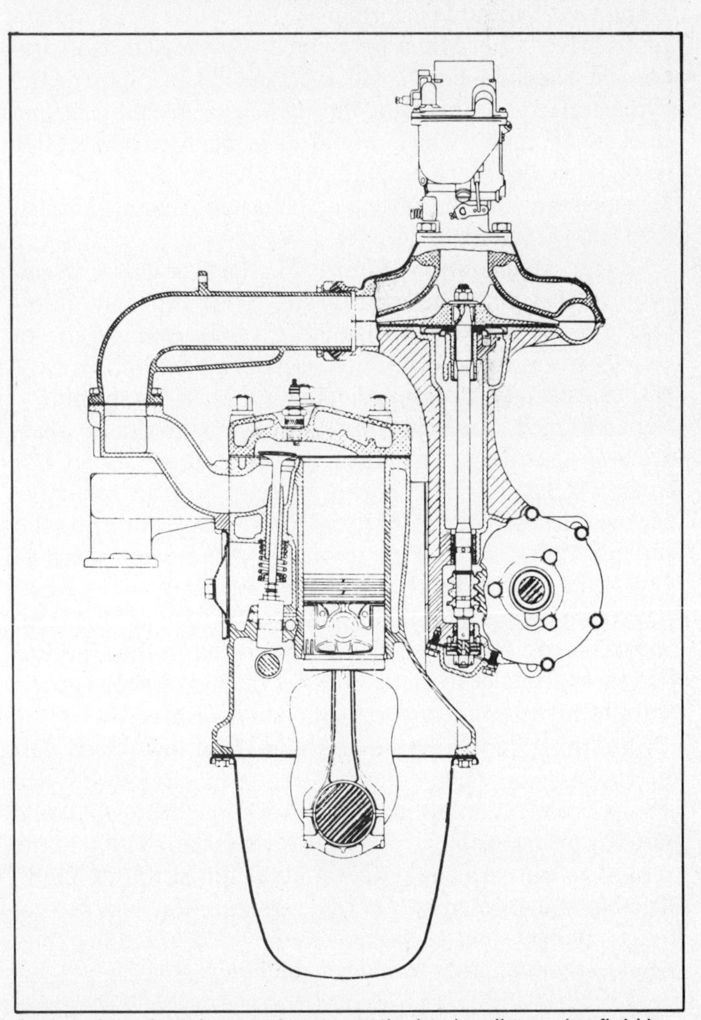

Graham introduced superchargers to the low/medium-price field in 1934. Gear-driven blower boosted the power of their eight-cylinder engine from 100 to 135 HP. Top speed was more than 90 mph, with good fuel economy.

Chrysler was the first large American automaker to mass-produce cars with improved body aerodynamics. On their Chrysler and DeSoto Airflow models of 1934, the streamlining added 5—10 mph to top speed. Nevertheless, the controversial styling didn't sell well.

clutched into 2nd gear to slow for a 30-mph corner! He did this for 24 laps without clutch or transmission trouble.

Hudson Terraplane-8—Did you know there were copiers of Ford-V8 low-priced performance in the mid-'30s? Only one year after the V8 was introduced, the Hudson Motor Company put their small in-line eight-cylinder engine in the lightweight Terraplane model. The package provided 92 HP in an under-3000-lb car for $650. Factory acceleration figures quoted a 0—50-mph time of 13 seconds and 0—75 in 21 seconds.

Hudson keyed their whole advertising theme for the Terraplane-8 toward performance. They even hired Indianapolis driver Chet Miller for a public top-speed demonstration on Daytona Beach. He averaged 85.84 mph for the flying mile. Then he went through again in 2nd gear and clocked 69 mph—which would have been close to 5000 rpm! The British magazine AUTOCAR said that the Terraplane-8 was the fastest-accelerating passenger sedan they had yet tested.

Yes, it could outrun a Ford V8. But it a was a *sleeper* even in that day. The man on the street probably didn't read the ads. The cars had little recognition. And, of course, it was the middle of the Great Depression. Few car enthusiasts today ever heard of the '33—'34 Terraplane-8.

Supercharged Graham—In 1934, Graham put a gear-driven centrifugal supercharger on their 265-CID, straight-8 engine. The blown Graham was an attractive package: 135 HP in a 3600-lb car for $1100. By utilizing the supercharger's turbulence to ensure more-even distribution of the fuel to the cylinders, they were able to run especially lean carburetion and get unusually high gas mileage. Supercharged Grahams often finished well in the Gilmore Economy Runs of the late '30s. Yet they gave peppy acceleration and a top speed near 90 mph in overdrive.

One thing should be noted about these low-priced performance cars. Their long suit was a quick getaway and strong acceleration up to 30, 40 or 50 mph. You still could not buy a car capable of 100 mph for less than $1000 in the mid-'30s. But you could have a lot of fun at the stoplight. The big high-horsepower luxury cars were too heavy to get away quickly—even though their good-breathing big-displacement engines would win out above 60 mph.

THE FIRST POWER PACKS

A *power pack,* as we understand it today, could be defined as a simple package of special engine equipment

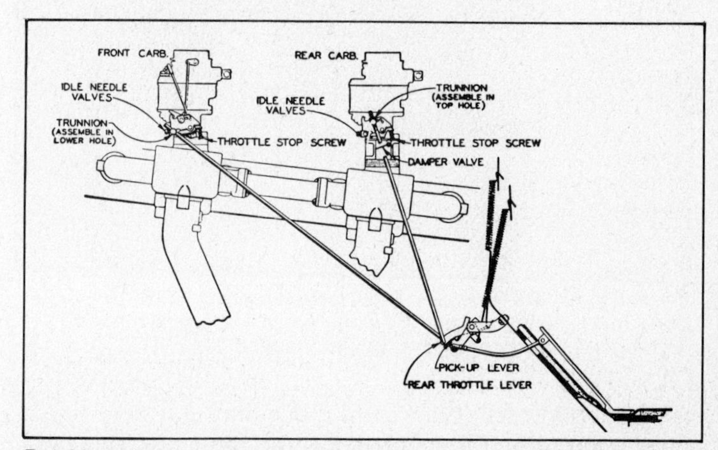

Buick's optional *compound carburetion* system of 1941 used two Stromberg downdraft carburetors on a log manifold. Engine cruised on front carb only. A damper valve kept rear carb closed until 80 or 90 mph. Engine developed 165 HP at 3800 rpm from 320 cubic inches.

the manufacturer offers to improve the acceleration of its cars. In most cases, it can be installed on the factory assembly line—though sometimes it must be installed by the new-car dealer. The pack might be nothing more than high-compression cylinder head/s or an oversize carburetor or multiple carburetors. It might be as simple as a dual exhaust system. The main idea is that a power pack is simple, bolt-on equipment that doesn't require complicated engineering, tooling or installation.

Detroit first started talking about optional power packs in the 1950s. But if you look back, the idea started long before that. Remember the Denver aluminum cylinder heads for the early Ford V8? A dealer-installed option, they provided a then-astronomical 7:1 compression ratio.

Several other companies, including Chrysler, Studebaker and Hudson, offered similar high-compression heads. For example, in the mid-'30s, Chrysler rated their big eight-cylinder engine 10-HP higher (140 HP) with the optional 7.5:1 head. It was said to give a measurable improvement in both acceleration and fuel economy. Premium gas in those days had a Motor octane rating of about 80, so it would allow around 7:1 compression without detonation problems. Most manufacturers used around 6:1 on standard engines to allow using regular-grade gas.

Buick went a step further in 1941 with their optional *Compound Carburetion* system. They used two downdraft, two-barrel Strombergs on a log-type manifold, with a progressive throttle linkage. The progressive linkage opened

Popular Ford Model A conversion in '30s was the 12-valve Riley. F-head used two side intake valves and one exhaust valve in the block. Riley conversions were made from '28 through '34 by George Riley. Photo by Tom Monroe.

Another modern-day Riley Ford nearly a half-century later. This one is the more-desirable four-port type. Photo by Tom Monroe.

the rear carb after the front one was about two-thirds open. Thus, you could cruise on just the front two barrels up to about 75 mph. Pushing the accelerator to the floor would whomp the rear carb open and give you a real kick. The system boosted the output of the big straight-8 from 150 to 165 HP. This assured a genuine 100-mph top speed. Acceleration was helped noticeably, too. Usually conservative Buick advertising people excused the gross indecency with references to "quicker response for city and highway driving."

The power-pack idea was no big thing in the '30s and '40s. But the point is that American car manufacturers were even then well aware of the performance-improving potential of a few simple pieces of external engine equipment. The lesson was not forgotten in later years.

THE CALIFORNIA REVOLUTION

The classic hot-rod sport that emerged in California following World War II was probably the most-important ingredient in triggering the performance carnival of the '50s and '60s. The sport was based on relatively crude $500 home-built hybrids—not a whole lot different from the Ford Model T speedsters of the World War I period. These classic hot rods used early Ford chassis and bodies. Most popular were the Model A and '32—'33 roadster bodies, usually with a highly modified flathead V8. Often updated with hydraulic brakes, they were always stripped of fenders and add-ons.

If there was an essential difference from the earlier Model T craze, it was that these modern hot rods appealed to a generally younger audience. You could get a driver's license at age 14 in most states, and high-school kids were more than ready for their first rod.

In some ways, the U.S. was quite affluent then. Kids had more *real* money to spend on fun. Add to this a flood of young soldiers returning from the war, with a new knowledge and appreciation of mechanical things. There existed all the ingredients for a tremendous boom in the auto sport. The fact that the participants were very young only

Hot-rod sport flourished following World War II. It was based on modifying the flathead Ford V8 engine. Outputs in excess of 200 HP at 4500 rpm were possible with a "full-house" of bolt-on goodies. Photo by Wally Chandler.

Hot street setup for the flathead Ford used two Chandler-Groves 97s on a close-coupled aluminum two-plane manifold—to allow room for the stock generator. Remember?

Advent of organized drag racing in the late '40s and early '50s taught enthusiasts a lot about acceleration performance. This whetted the youth market for the factory supercars that were to come later. Photo by Wally Chandler.

increased the impact and solidity of the trend. The kids weren't kidding. They wanted cars *and performance*.

It's hard to pinpoint the real roots of the modern hot-rod sport. There was certainly important activity before World War II. A hard core of enthusiasts were racing stripped Fords for top speed on the California dry lakes in the late '30s. Pioneers like Vic Edelbrock and Eddie Edmunds were making bolt-on speed equipment before the war. This period also saw the beginning of more-sophisticated bolt-on goodies to help performance. Some of these items were sold in traditional auto-parts stores. Easily available were things like the Columbia two-speed rear axle for Fords, or the McCulloch belt-driven supercharger.

So, World War II was an interruption in a movement that was going to happen anyway. It fueled the fire.

Things really started to move after the war. The economy boomed. That meant jobs for young people, and more money to spend on this new car hobby. A whole industry quickly sprang up to supply special equipment to hop-up the Ford flathead V8. By 1950, you could buy anything from special high-compression cylinder heads and multiple-carburetor manifolds to high-lift, long-duration cams, special pistons, dual-coil distributors, stroker crankshafts and welded, tubular-steel exhaust headers.

In addition to the engine parts, an increasing variety of special chassis and body equipment became available. Remember the early Kinmont disc brakes to fit the Ford spindle? Or how about the DuVall "V" windshield to dress up the '32 roadster cowl? The DuVall was styled after Beuhrig's windshield for the '35 Auburn Speedster!

Another phenomenon developed—the *speed shop*. In the early '50s, speed shops began popping up in major cities all over the U.S. You could buy all sorts of high-performance goodies over-the-counter. One of the first was Roy Richter's Bell Auto Parts in Bell, California.

The growing demand from grass-roots hot rodders was sufficient to support a significant specialty industry on the West Coast. Performance grew to a multimillion-dollar business within a few years after World War II. And this was entirely apart from any performance-oriented engineering development that might have been going on in Detroit at the time. Detroit didn't recognize the do-it-yourself performance sport until much later.

Many supercar fans tend to downgrade the performance of those early home-built hot rods. Don't you do it too. A postwar Ford flathead block could be readily bored and stroked to 300 CID. With a good porting and relief job, airflow through the side valves into the bores was improved. In addition, all the usual bolt-on goodies such as heads, manifold, cam, domed pistons, ignition, headers and so on, could easily double the output. Numerous dynamometer tests indicated 200—220 HP at 5000 rpm on pump gas. With total car weights around 2000 lb, acceleration was as good as most modern supercars. How do 0—60-mph times around 6 seconds and top speeds of 130—140 mph grab you? Few, if any, production cars in the world could top the straight-line performance of a California hot rod in the early '50s.

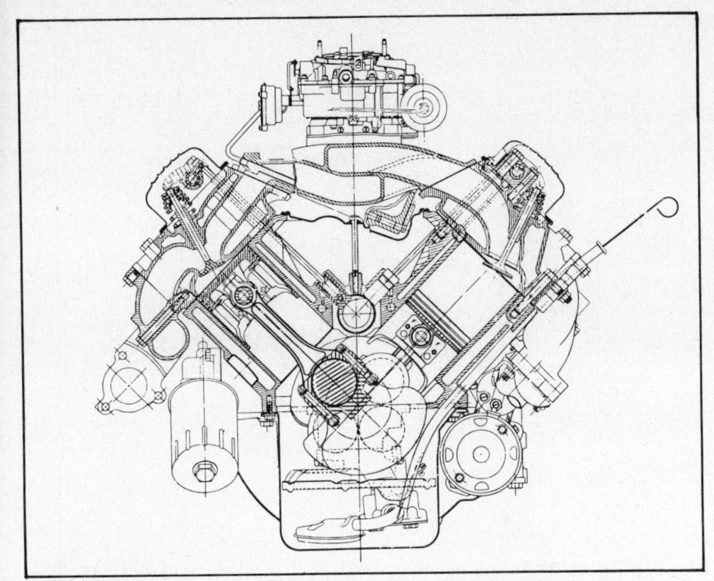

General Motors introduced the modern overhead-valve, short-stroke V8 engine in Cadillacs and Oldsmobiles in 1949. It revolutionized stock-car performance overnight. This is the early Olds Rocket V8. Drawing courtesy of General Motors.

DRAG RACING HELPS OUT

Perhaps it was inevitable that the brute performance of these simple home-built cars would spawn a new type of competition. This competition emphasized acceleration rather than flat-out speed. They called it *drag racing*—side-by-side, straight-line acceleration from a standing start over a quarter-mile course.

At first, drag racing was not a whole lot different than the informal stoplight racing first seen in the '30s with V8 Fords. But when organized with safety rules, electric timing and a class structure for different types of cars, the new sport caught on quickly. The first informal races were held on abandoned California airstrips at Santa Ana and Goleta, California in the late '40s. Then, in 1951 the National Hot Rod Association (NHRA) was formed. In 1955, the NHRA held its first National Championship meet at Great Bend, Kansas. The sport had national recognition overnight.

But drag racing proved to be more than just a fun sport in the evolution of automotive performance. It was a form of competition where any car fan could participate with minimum investment. The growing market encouraged rapid proliferation of special equipment to help the drag-strip performance of hot rods and standard cars. This not only included engine-power goodies, but special axle gears, high-traction tires, traction bars, shifters, tachometers, aluminum flywheels, special clutches and so on. With the hardware came a whole new technological discipline on how to apply it for better acceleration.

No one can ever be sure how much drag racing precipitated the American supercar. But, personally, I don't think we would have ever had supercars without drag racing.

A NEW TYPE OF ENGINE

At the time the early hot rodders were refining the Ford flathead V8, there hadn't been anything really new in American automotive engines for 20 years. In the late '40s, most Detroit engines were in-line sixes or eights, with side valves or simple pushrod/overhead-valve layouts. These engines featured relatively long strokes in relation to bore—decidedly *undersquare*. Detroit had been making engines that way since the early days of the automobile. Why change?

We should credit a handful of engineers in General Motors' research department, headed by Charles "Boss" Kettering, for changing things. Actually, all they wanted to do was to find a way to use higher compression ratios with available pump gas. But when they dug into the problem, they had to design a completely new type of engine to run 1.5-points higher compression than the existing engines. What came out was the modern, short-stroke, overhead-valve V8. And it revolutionized passenger-car performance overnight.

Why the V8? Actually, it's simple. When GM researchers tried to use higher compression with in-line engines, they got so much torsional vibration in the long crankshafts that the engines ran too roughly. A short four-throw crank helped a lot. But even then there was a lot of whip. So the engineers shortened the stroke to get more crankpin overlap and a stiffer crank. They enlarged the bore to keep the cubic inches. And of course they had to use the "V" cylinder arrangement to get eight cylinders with a four-throw crankshaft. The overhead valves were necessary to get a compact combustion chamber and short flame travel, to allow maximum compression on a given fuel octane.

You can see they did it all for compression. But the side benefits were actually more important to performance. Overhead valves gave better porting and breathing. A bigger bore made room for larger valves. The V8 cylinder arrangement gave a good intake-manifold layout for more-even air/fuel distribution. The short stroke reduced internal friction, and increased net horsepower for the amount of fuel burned. The more-compact cylinder block reduced overall engine weight.

Everything seemed to work together—almost better than the GM folks expected. Results were so good in the lab that tooling was started shortly after the war for brand-new, short-stroke V8s for the 1949 Cadillacs and Oldsmobiles. These engines could use 7.5:1 compression with the currently available 80-octane, regular-grade gas. But just as important: The V8s developed more horsepower and torque per cubic inch and per pound of weight than the old-style long-stroke, in-line engines. They also gave better fuel economy, longer life, more revving ability, less noise and less vibration—everything that counted in a passenger-car engine.

Imagine the performance improvement. The lightweight '49 Olds 88 with the *Rocket V8* could turn top speeds near 100 mph, and 0—60-mph times of 13 seconds. Combined with unheard-of throttle response, smoothness and 18—20-mpg fuel economy, the car was a hit in Oldsmobile showrooms.

Overnight, Detroit had a whole new performance future to play with.

One of the first high-performance cars out of post-World War II Detroit was the beautiful Chrysler 300. These were big, heavy cars, yet very powerful and good handling. Photo by Ron Sessions.

Power Packs in the '50s

Muscle for the masses

NASCAR CHANGES THE PICTURE

Organized forms of auto racing have always had a positive effect on the performance development of standard passenger cars. Fender-to-fender competition seems to make the fans think more about the performance of their own cars. This is especially true of stock-car racing, where the fans identify with specific makes and models on the track.

Stock-car racing never really came of age in America until the formation of the National Association for Stock-Car Auto Racing (NASCAR). In a little motel in Daytona Beach, Florida in December, 1947, NASCAR was born. Under the powerful leadership of Bill France and a sharp crew of promoters and technical people, the infant organization quickly grew. Soon, it challenged the top open-wheel, oval-track organizations in the country—including American Automobile Association (AAA) Championship racing.

By the mid-'50s, the NASCAR Grand National carnival was playing on dirt tracks and banked speedways across the South. In 1958, France built the magnificent 2-1/2-mile oval Daytona International Speedway. Daytona boasted seating for 60,000. The high-banked track was capable of handling speeds up to 200 mph. It was stock-car racing's Indianapolis. There was no stopping NASCAR then.

Above: Small-block '55—'57 Chevrolets were quick with the optional 283-CID Corvette engine. The '57 model shown has clean, uncluttered lines. Photo by Ron Sessions.

Certainly, the most-direct effect NASCAR racing has had on passenger-car performance development was hidden in their rule structure. Early Grand National class rules required practically all equipment on the cars to be stock. This included almost all engine components: heads, manifolds, camshaft, pistons, carburetors, ignition. It also included most chassis parts and running gear. You couldn't even use a set of tubular-steel exhaust headers in those days. The rules required factory cast-iron exhaust manifolds with factory part numbers. You could, however, run an open pipe off the manifold.

Needless to say, an auto company wanting to look good on the NASCAR tracks had to get busy and do some performance development in the lab and dyno rooms. However, there was some flexibility in the rules. The resulting special equipment did not have to be offered as an assembly-line option. It could be distributed strictly over-the-counter. But the rules said it had to be available to the public, with factory part numbers. No special parts for special people—they hoped.

Admittedly, some companies had parts made for them by established hot-rod suppliers. Ed Winfield and Ed Iskenderian were the top names in high-performance camshafts in the early '50s. So it was natural for inexperienced factory engineers to go to them for help. For example, the Iskenderian E-2 camshaft grind was used by several companies for racing. Isky ground the cams on factory billets, stamped them with a factory part number, and that was it.

Back when Richard Petty was knee-high to a NASCAR draft, his father, Lee Petty was winning stock-car races in a big Chrysler. Lee brought home the bacon in this '53 Daytona Beach race. Photo courtesy of Daytona International Speedway.

Stock-car race cars were *really stock* in the '50s. This is the dashboard of Herb Thomas' famous 1952 Hudson Hornet. It even had the heater hooked up—and the clock worked! Photo by Wally Chandler.

It was perfectly legal as long as you could buy that cam through a new-car dealer. Isky also supplied high-rate valve springs and heavy-duty retainers.

Incidentally, special racing parts were first referred to as *export* equipment in the early '50s. The Detroit car companies were still pretty skittish about out-and-out involvement in racing. They feared criticism from government officials and safety people. No need to rock the boat. The factories were only interested in getting new equipment to professional racers anyway—not to the man on the street. It was simple and convenient to disguise the whole program under the export label. Many car enthusiasts were entirely aware of what was going on.

RACERS UNDER THE SKIN

Certainly the most-active car company in early-'50s NASCAR racing was Hudson. They introduced their unique *step-down* chassis in 1949. The Hudson step-down used a perimeter frame with semi-unit shell construction. You stepped over the frame rail when entering the door. This allowed car height to be reduced, lowering the center of gravity (c.g.). Careful matching of springs and shocks, and the lower c.g., resulted in outstanding handling—for those days. Also, Hudson's big 308-CID flathead six was a fairly healthy engine. And the cars were not overly heavy. It seemed natural to promote their good performance and handling through an aggressive NASCAR program.

Marshall Teague was the top Hudson pilot in the early '50s. "Step-Down" Hudsons featured a perimeter frame that went outboard of the rear wheels, enabling the driver to lay the right-rear fender against the Armco and guide the car through the turns without damaging the right rear tire. By the way, this photo is courtesy of Vince Piggins. Thanks Vince!

HUDSON WANTS TO BE YOUR STOCK-CAR COMPANY

When Hudson introduced their *step-down* chassis cars in 1949, it suprised a lot of people—including mighty General Motors. GM immediately went out and bought a number of the new Hudsons to scrutinize. The step-down cars were used as a standard of handling and performance at the GM Proving Grounds for several years.

Stock-car racers who ran the beach at Daytona knew the flathead-six Hudsons were good runners too. A young Smokey Yunick was building Hudson engines and cars, and a Pure Oil gas-station operator by the name of Marshall Teague was winning with them. Still, it was an independent effort.

Bill France and other supporters had just founded the NASCAR organization. They were interested in stirring up some factory support to really get things moving. France took the enthusiastic Teague up to Detroit to sell Hudson on the idea of supporting a racing team. Seeing an opportunity to forge a performance image in an increasingly competitive new-car market, Hudson went for it.

The only problem was putting together the list of "stock" parts to ensure Hudson stayed out in front. The Hudson people put the young engineer, Vince Piggins, in charge of their newly formed Severe-Usage Parts Program.

All the stock-car tracks were dirt back then—very tough on parts. In cooperation with Kelsey-Hayes, Piggins and company developed the first reinforced wheel. Up to then, racers had been reinforcing the spiders with boiler plate—which tended to break spindles. Other Severe-Usage Parts the Piggins group developed included larger-diameter rear-axle shafts, shot-peened spindles and steering arms and much more.

Hudson Hornets were holding their own against the Oldsmobile Rocket V8s on the circle tracks in the early '50s. Then Olds introduced 4-barrel carburetion for 1952, giving it a performance edge. Hudson countered with the now-famous 7X engine.

Introduced at the Detroit Fairgrounds that year, the 7X put Hudson out in front again. As Vince Piggins tells it, the flathead-six 7X underwent a special machining operation whereby the block was *plunge-cut* to unshround the valves. This improved the engine's breathing, especially at higher revs. Plunge-cutting was similar to the relief job the racers gave the Ford flathead V8s to improve their breathing.

Another feature of Hudson's step-down perimeter-frame design gave the cars an edge in "handling." The frame rails were *outboard* of the rear wheels! This enabled Marshall Teague and the Herb Thomas Team Hudsons to employ a unique driving technique.

The Hudson pilots would actually lay the right-rear fender against the Armco rails and nail the throttle. Where the other stockers had to go low in the turns, the fastest line for the Hudsons was high and wide! Sparks would fly and the cars would be gouged down the right side, but they won races that way. The resulting tortured bodywork became known as *Darlington Stripes,* named after the track where the technique was most-often practiced.

Marshall Teague at Daytona Beach. Masking tape reduced damage from flying sand! Bungee cords kept the hood shut. Photo courtesy of Daytona International Speedway.

One factor that limited top speeds of passenger cars in the '50s was tires. They weren't designed for 100-mph speeds. This one peeled its tread after six miles at 115 mph. Photo by Wally Chandler.

Hudson engineers developed a special-parts list as long as your arm. The list included not only engine equipment, but heavy-duty springs, shocks, hubs, spindles, wheels, frame reinforcements—and even special axle and transmission gears. You could build a raceworthy Hudson using all factory parts.

Special parts for the 308-CID engine included an aluminum high-compression head, dual carburetion, high-performance camshaft, high-rate valve springs, heavy-duty pistons, rods and oil pump, and dual low-restriction exhaust manifolds. Hudson called it the *7X* engine. It was said to develop about 210 HP at 4500 rpm. Doesn't sound like much today—but it was enough to let the Herb Thomas Team dominate NASCAR racing in the '52—'54 period.

The only Hudson racing hardware to reach the assembly line was the dual-carburetion system, called the *Twin H-Power* option. The big six was rated 170 HP at 4000 rpm with the system. It was enough to push the popular Hudson Hornet to a genuine 100-mph top speed, with 0—60-mph times around 12 seconds. A dual-carb Hornet would definitely outrun a V8 Olds 88 in those days. The biggest problem was the Hudson's long 4-1/2-in. stroke. This limited peak revs to 5200 rpm. It took a lot of development at the Hudson factory to make this obsolete bottom end reliable for the long NASCAR races.

I might mention that Oldsmobile dominated NASCAR racing in the '50—'51 period, with benefit of their new short-stroke, overhead-valve V8. Race officials made them use the stock hydraulic valve lifters. Pit mechanics had a terrible time keeping the lifters from pumping up at over 4000 rpm. Another handicap was that Olds only had 2-barrel carburetion. Their two-year domination of stock-car racing with these technical handicaps says a lot for the basic performance potential of the new Kettering engine.

Dual-carb Hudson Hornets and V8 Olds 88s were the hottest American production cars on the road in the early '50s. Early Chrysler Hemi V8s could challenge them for top speed, but Chryslers were too heavy for quick acceleration. After Olds and Cadillac got 4-barrel carburetion in 1952, they temporarily moved to the front of the performance race. But the dual-carb Hornets could hang right in there. All Detroit automakers offered 4-barrel-carburetor engines within three years.

And the infamous horsepower race of the 1950s was on.

AN ENGINE THAT CHANGED THE WORLD

I mentioned earlier that the GM/Kettering engine concept of the late '40s—short stroke, overhead valves, high-compression V8 layout—had a crucial impact on the engineering evolution of the American passenger car. Within five years, the entire industry converted to this type of power.

No less a milestone on the performance trail was the small-block Chevrolet V8 of 1955. This engine literally changed the world. It broke new ground in almost every area of design. It had the smallest dimensions and lightest weight per cubic inch of any engine in the industry. It used new *thinwall* casting techniques to save weight and cost. It used fewer and simpler casting cores than any previous design. A simplified valve train featured stamped rocker arms pivoting on stud-mounted ball joints. Lubrication was fed to the valve train through lightweight, hollow pushrods. Hardly anything about the new Chevrolet engine was like other V8s of the period.

But perhaps the most-remarkable thing about the new engine was its unusual *performance*. I'm sure the designers were as pleasantly surprised by this as anyone else. From

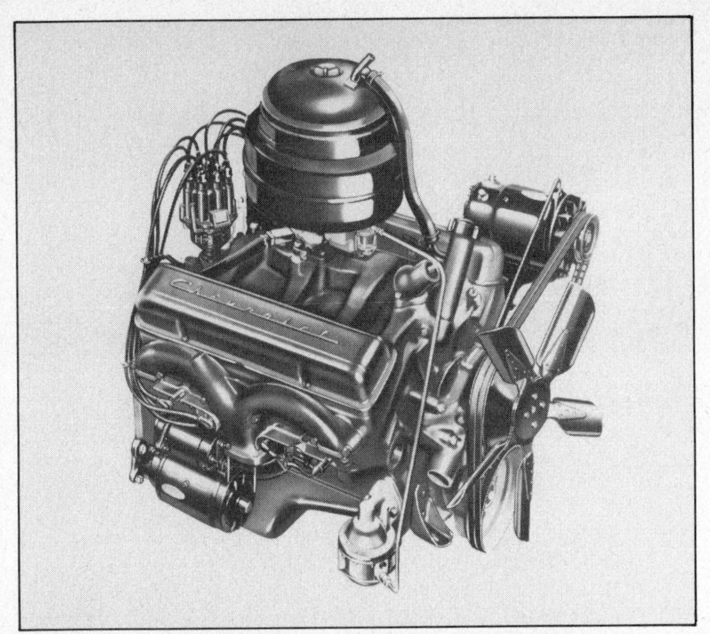

Chevrolet's new small-block V8 revolutionized passenger-car performance in '55. It weighed only 530 lb, gave excellent fuel economy, and developed more horsepower than most big-blocks. Photo courtesy of Chevrolet.

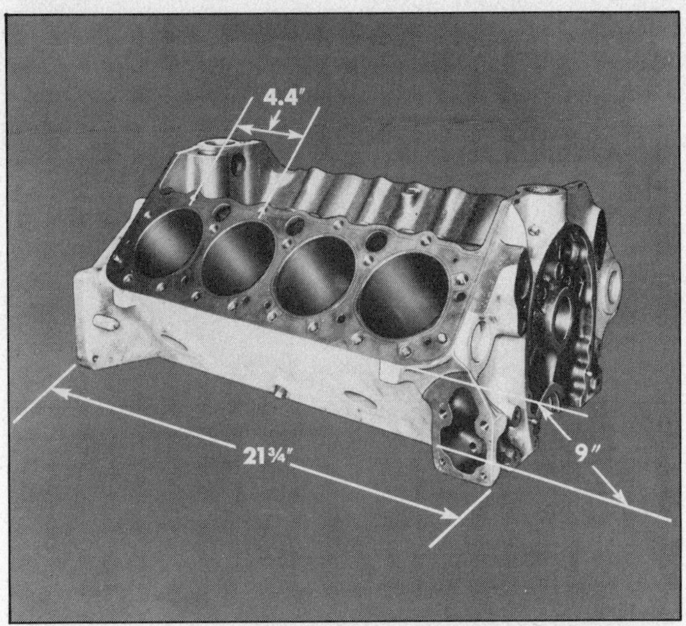

Cylinder block of the new Chevy V8 featured *thinwall* casting techniques, low deck height, and weighed only 145 lb complete. Low weight saved cost and helped performance. Photo courtesy of Chevrolet.

the first day that new configuration went on the dynamometer, it did things that no bread-and-butter passenger-car engine displacing 265 cubic inches had any right to do. It developed 180 HP at 4800 rpm with 8:1 compression and a 4-barrel carburetor. It had less friction at 4000 rpm than any engine in the industry. It showed less heat rejection to the coolant, indicating high thermal efficiency. This overall efficiency was supported by low fuel consumption. The light *crank train*—crankshaft, rods and pistons—gave extremely quick throttle response; an important factor in acceleration.

And, perhaps most important, the ultralight valve train permitted the engine to turn 5500 rpm or more with stock valve springs and hydraulic lifters! This is what sold the engine to the performance buffs. All the other V8s had to be shifted at 5000 rpm or less to avoid valve float and lifter pump-up—too close to the power peak for optimum through-the-gears acceleration. By contrast, the new 265-CID Chevy could wind almost 1000 rpm above the stock power peak—a tremendous advantage in drag racing.

As I said, the unusual performance and efficiency of the new Chevy V8 was probably as much a surprise to the designers as to the car buyers. Everything seemed to work right. When you stepped on the gas, things happened—whether it had a 2-barrel or 4-barrel carb, single or dual exhaust. The engine was born to *run*.

It wasn't long before the Chevy's dynamometer performance was justified on the road and at drag strips. *ROAD & TRACK* Magazine tested a '55 two-door with the optional *power pack*—4-barrel carb and dual exhausts—with 4.11:1 final drive and overdrive. They clocked a 0—60-mph time of 9.7 seconds. Standing-start quarter-mile elapsed time was 17.4 seconds, hitting 77 mph through the lights. Although the engine was capable of

Chevrolet V8s of the '55—'57 period were the supercars of their day. Chevy stole the youth market from Ford in a few short years. Photo by Ron Sessions.

Light Chevrolet coupes of '55—'57 period were the scourge of the drag strips in the late '50s. Many of those raced had the optional Corvette engine. Photo by Wally Chandler.

A '57 Chevrolet V8 convertible in its element, parked outside a high school. 1957 was a watershed year for early Chevrolet V8s—a bore change bumped displacement to 283 cubic inches. A pair of fuel-injected engine options were available in passenger cars that year, the top one producing 283 HP at 6200 rpm. No, the fender skirts were not stock.

revving to 5600, shifts were made at 5200 rpm with the three-speed column shift.

R&T also tested for top speed, timing a two-way average of 104.7 mph in direct-drive high—with the engine turning 5400 rpm. The car was only 2-mph slower in overdrive, turning 3000 rpm. This indicates a high peak on the net-power curve, as installed in the car. According to Chevy engineers, this was 160 HP at 4400 rpm for the 4-barrel engine with dual exhaust.

Anyway you measure it, this was revolutionary performance in 1955. No other American passenger car could touch the small-block Chevy's acceleration. Suddenly, power-pack Chevys were the cars to beat at the stoplight and on the drag strips. Before the small-block, Chevy had no reputation for performance. Up to then, the quickies were the Olds 88s, Hudson Hornets, light Cadillacs, and even some flathead Fords. But this was an entirely new role for the company known for its lowly "stovebolt" six.

THE POWER-PACK SYNDROME

I mentioned in the first chapter that the idea of a *power-pack*—a package of optional engine equipment to improve performance—started as far back as the mid-'30s. Chevrolet's 4-barrel carburetor and dual-exhaust package of 1955 was a simple extension of the idea. But when these goodies were added to an engine that already had more basic performance potential than any Detroit engine up to that time, the power-pack idea took on an entirely new meaning. All of Detroit's automakers jumped on the bandwagon within two years.

Those power packs of the mid-'50s followed a similar theme: multiple carburetors, a high-performance camshaft and a simple dual exhaust system. Carburetion was either two 4-barrels or three 2-barrels.

In the mid-'50s, 4-barrel carburetors were in their infancy. They didn't have very efficient breathing. Until Holley started rating its carburetors by airflow capacity in '57, most carburetor manufacturers listed their carbs by venturi size. But if those early 4-barrels had been tested on modern flow rigs and rated by modern standards, they would have shown flow ratings of only 400—450 cu-ft per minute (cfm). Most mid-'50s 2-barrel carbs flowed only 200—250 cfm.

Obviously, mid-'50s V8s really *needed* two 4-barrels or three 2-barrels for sufficient breathing.

Camming was equally conservative in those days. Engine designers were fully aware that higher valve lifts and longer durations could help top-end horsepower. But they didn't know much about refining cam-lobe profiles to give better mid-range torque at the same time. Those mid-'50s cams ran relatively strong on the top end, but they usually had a lumpy idle and poor low-rpm response.

Dual exhaust systems were used on the theory that doubling the passage area of a gas-flow system would reduce the restriction by 75%. It worked well enough. But those early low-restriction exhaust systems would have been more efficient if the designers had included special low-restriction exhaust-manifold castings as well. These refinements were to come later.

Here's a brief rundown of the design concepts of those early power packs in the individual companies:

Chevrolet—After the unexpected success of the first 1955 power pack—which was just a 4-barrel carb and dual exhausts—it was a natural step to move to two Carter WCFB

Distinctive split grille of Ed Francis' '56 Chrysler 300-B. Photo by Ron Sessions.

In '56, the Chrysler 300-B added a first place victory in the Daytona High Performance Trials to previous AAA and NASCAR trophies. Photo by Ron Sessions.

4-barrels and a high-lift cam for '56. This was the top Corvette engine that year, and it was offered as an option in passenger cars the following year. The package also included special finned aluminum valve covers with the name CORVETTE cast in. Small individual air cleaners were used. Also, an elaborate progressive throttle linkage let the car cruise on only the two front barrels of the front carb.

Chrysler—The Chrysler Hemi was introduced in 1951. That year, with modest 7.5:1 compression, the 331-CID monster developed 180 HP. Chrysler's first real power pack was for the *Model C-300* sports/luxury coupe in 1955. It consisted of dual 4-barrel carburetors, a long-duration, solid-lifter cam, and dual exhausts for the big 331-CID Hemi engine. Advertised rating in '55 was 300 HP at 5200 rpm, which is how the *300* got its name. The power and torque of this unusual hemi-head engine could move the 4600-lb Chrysler around in brisk fashion. With 0—60-mph times in the 9-second bracket and a top speed near 130 mph, the big Chrysler suddenly became America's best-performing production luxury car.

In 1956, further development bumped output of the Chrysler 300B's 354-CID Hemi to 340 HP. An optional engine package available for the big 354-CID Chrysler was rated at an astounding 355 HP. This was a milestone for an American production-car engine—more than 1-HP per cubic inch! Eventually, displacement was increased to 392 cubic inches and output to 390 HP.

The formula worked so well, in fact, that Chrysler later offered a 300-type sports coupe in each of its lower-priced car lines. These were the mid- to late-'50s DeSoto Adventurers and Dodge models with the D-500 engine-option package. The early models used small-block Hemi engines of various displacements in the 241—345 cubic-inch range. They were patterned after the big Chrysler and used the same type of power equipment—dual 4-barrels, and so on. They were almost as quick as the big 300.

Plymouth didn't get left out either. Though Plymouth didn't get to use the Hemi, it offered a Fury coupe with a high-performance V8. This was the A-series engine with *polyspherical*-combustion-chamber heads. The '57 Fury

Chrysler used dual 4-barrel carbs, high-performance cam and dual exhaust to get over 300 HP from the early 300s. Even with more than 4500 lb of weight, the 300 had brisk acceleration and a 130-mph top speed. Photo by Ron Sessions.

With luxurious interiors gracing the car's outstanding handling and performance, Chrysler 300s were considered grand touring cars by many. Photo by Ron Sessions.

1958 Chrysler 300-D was a car of massive proportions, yet managed to come off gracefully. Chrysler advertised the car as the "perfect example of modern architecture." Big 392-CID Hemi was available in '57—'58 models. Photo by David Gooley.

Chrysler 300 was the first U.S. car to use a pleated-paper air filter. Up to then, oil-bath air cleaners were the rule. An excellent example of racing improving the breed, Chrysler found in their mid-'50s stock cars that oil in the old-style air cleaner would surge to the side of the housing on hard turns, leaving the engine unprotected. Photo by David Gooley.

was a potent performer with a 290-HP 318-CID engine with dual 4-barrels.

Ford—Ford disguised their early power packs as optional *Police Interceptor* engines. The first one in 1955 was a standard 292-CID 4-barrel with a special solid-lifter cam, high-rate valve springs and dual exhaust. No specific HP rating was put on this package. In '57, Ford went to 312 CID and two 4-barrel carbs for the Police Interceptor package. This time they rated it 285 HP at 5000 rpm. Its dual 4-barrel carburetion system had been used for NASCAR racing in '56, but it was not offered as an assembly-line option. Later in the '57 model year, Ford developed a supercharger package for NASCAR racing with the 312 engine. More on that engine later.

Oldsmobile/Pontiac—These GM divisions took a little different tack with their first power packs in 1957. Instead of two 4-barrel carburetors, they chose the three 2-barrel route. They reasoned that the 6-barrel grouping along the manifold would more-evenly distribute the air/fuel mixture—especially when cruising on the center carb. By contrast, most of the dual-quad systems cruised on the front carb only. This gave uneven distribution at small throttle openings. The Olds and Pontiac people also liked the looks of a three-carb setup under the hood. They felt their *trips* were definitely more impressive than two *quads*. Appearance was always a vital ingredient in power-pack design!

Incidentally, the Olds and Pontiac sales people went to some effort to think up catchy identification for their new multiple-carburetor engines. Olds used the *J-2* label on their hood. Pontiac dug up the *Tri-Power* name, which they stuck with for another ten years. Both engines also came with the usual long-duration cam—hydraulic lifters this time instead of solid—and low-restriction, dual exhaust systems.

Oldsmobile, Pontiac and Chevrolet opted for triple 2-barrel carburetion for their big-displacement performance engines in the late '50s. This is the famous *J-2* setup on the Olds Rocket V8, with vacuum-operated, progressive throttle linkage.

Model	CID	Adv. HP @ rpm	0—60 mph	1/4-Mile e.t. @ mph	Top Speed
'56 Chevrolet, Corvette eng.	265	225 @ 5200	8.7 sec	16.3 @ 85	123
'57 Dodge D-500	325	310 @ 4800	8.3	15.8 @ 88	NA
'57 Ford Interceptor	312	285 @ 5000	9.3	16.8 @ 83	119
'57 Oldsmobile 88 J-2	371	312 @ 4600	8.7	16.2 @ 86	122
'57 Pontiac Tri-Power	347	290 @ 5000	8.6	16.4 @ 85	NA

Most NASCAR racing was on crude dirt tracks in the '50s. This is part of the beach-road course at Daytona Beach. Appearances notwithstanding, dirt-track stock-car racing really helped get Detroit started in the performance business.

Factory sponsorship really heated-up stock-car racing in the mid-'50s. Here's some action out of the south turn at Daytona in 1955. Photo courtesy of Daytona International Speedway.

Buck Baker and his Kiekhaefer-prepared '55 Chrysler 300. Photo courtesy of Daytona International Speedway.

The Flock Brothers and their victorious '55 Chrysler 300s. Chrysler won all the NASCAR marbles in 1955. Photo courtesy of Daytona International Speedway.

None of these early power packs were a disappointment on performance. They gave big power boosts over corresponding standard production engines. I had the opportunity to road-test several of them back in the '50s. I borrowed them from car dealers or factory engineers. All were in showroom-stock trim, with no trick gear ratios, equipment or special tuning. At left are some 0—60-mph times, standing 1/4-mile e.t.s and terminal speeds. In a few cases, I was able to time a top-speed run as well.

DETROIT GOES RACING

In retrospect, it's hard to say whether the power-pack idea came before the idea to go racing, or whether it was the other way around. Anyway, the Detroit performance revolution exploded in the mid-'50s.

Everything seemed to happen at once. High horsepower ratings, true or not, suddenly became an important sales factor. The horsepower race was on. At the same time, superior top-speed potential and acceleration became a vital factor in the youth market. This triggered the introduction of power packs—which gave a big performance boost for only a few dollars. It didn't seem to matter that the power packs caused fuel economy and idle quality to suffer.

The renewed emphasis on performance naturally pushed the car companies into stock-car racing. In part, the factory racing programs were justified as a proving ground for developing performance equipment. But mostly, the programs helped to publicize and promote performance with the buying public.

In the '50s, NASCAR stock-car racing meant just that—*stock cars.* You could see a definite parallel between the performance of specific models and engine combinations in the mid-'50s, and their success on the NASCAR race tracks. For instance, in 1955 the big Chrysler 300 was the car to beat. It was anything but a promising track car, with its huge size and weight. But it had, by far, the most-usable horsepower of any American passenger car at that time. It translated directly into track performance, especially on the fast speedways.

Chrysler 300s continued to dominate in 1956, but Ford and Chevrolet began throwing a lot of factory development and support into the NASCAR picture. It wasn't long

Daytona Beach action in 1956—Joe Weatherly's Ford slides past "Speedy" Thompson's Mercury. Photo courtesy of Daytona International Speedway.

In the early days of NASCAR, there was a convertible racing class. This is Paul Goldsmith in the Smokey Yunick-prepared '56 Chevy. Photo courtesy of Daytona International Speedway.

Chevrolet jumped into NASCAR racing with both feet in 1956, after their new V8 proved its performance potential. Here's a row of 265-CID dual-4-barrel Corvette engines awaiting shipment to NASCAR factory teams.

Ford used the belt-driven Paxton supercharger on their 312-CID Thunderbird engine to try to beat the Chevys in 1957. NASCAR versions got 330 HP, and had the edge on the fast speedways. A few production versions were sold to the public. Photo by Wally Chandler.

before it paid off in short-track victories for the two. By 1957, Ford and Chevy were ruling the roost, pretty much dividing the Grand National spoils between them. Chrysler products were forced out by sheer cubic dollars. Some say Ford and Chevy were each spending $2—3 million a year on their NASCAR programs in the '56—'57 period.

Activity reached a peak early in the '57 season. Chevrolet adapted the new 283-CID Corvette engine with Rochester fuel injection to passenger cars. Ford countered with the popular belt-driven Paxton supercharger on their 312-CID V8 as an assembly-line-installed option. No one doubted that these moves were made primarily to help their cars on the NASCAR tracks. Only a handful of them were ever built and sold to the public. But they did make great racers. Both engines were capable of over 300 HP in racing trim with open exhaust. The supercharged Ford had the edge on fast speedways, while the injected Chevys were quicker on the shorter tracks. Between them, they were the fastest passenger cars money could buy in 1957.

Some of these blown Fords and injected Chevys made it into the public's hands. On the street, both cars were capable of 0—60-mph times in the 7-second bracket. Drag-strip quarter-mile times in the mid-15s at over 90-mph trap speed were commonplace. For an idea of top-speed capability, a blown Ford won its class in the '57 Daytona Beach flying-mile runs at 131 mph. An injected Chevy went 130 mph.

The Ax Falls—Two things happened in the middle of the 1957 racing season to change the whole NASCAR picture. For one, NASCAR officials began to get uneasy about the high cost of setting up a competitive Grand National stocker. Right out of a clear blue sky, they banned superchargers and fuel injection. All cars were required to use one 4-barrel carburetor.

But just as important as NASCAR's bombshell was a sudden, unexpected resolution from the Automobile Manufacturers Association (AMA) that took the Detroit

Chevrolet offered a big-block V8—the "W" engine—for the first time in 1958. The 315-HP high-performance version (348 CID) gave 15-second quarter-mile e.t.s in the full-size Impala. By 1961, the W engine had grown to 409 CID. Photo courtesy of Chevrolet.

car companies out of racing altogether. The ruling not only banned direct factory participation in racing, but it gently suggested the companies should stop emphasizing horsepower and performance in their advertising.

The AMA had a lot of reasons for the move. Ford and Chevy were spending millions on their NASCAR programs, really butting heads. Safety authorities were upset about the emphasis on speed and acceleration in advertising and sales promotion. Insurance people were scrutinizing rates on sports and high-horsepower models. And worst of all, government officials in the regulation and antitrust areas were watching the automakers closely. You might even say that some Feds were relishing the idea of the Detroit automakers knuckling under from their newfound regulatory zeal.

The obvious answer was for Detroit to take a lower profile on horsepower and performance. It did—for a while.

Corvette fuel-injection engine was available in passenger cars in 1957. Photo courtesy of Cars & Parts magazine.

SECOND-GENERATION POWER PACKS

The fuel-injected 283-CID Corvette engine of 1957 was a milestone in American automotive history for several reasons. It was the first U.S. production engine to use port fuel injection. It was one of the first U.S. production engines to claim an output of 1 HP-per-cubic inch of displacement. And it was the first U.S. production engine with a factory-approved rev range beyond 6000 rpm. In fact, the bottom end and valve train were capable of a then-astronomical 6500 rpm.

But the important thing about this engine from the standpoint of American-performance-car evolution is that it represented a second-generation power pack. The Rochester-injected Corvette was more than a bunch of bolt-on, easy-to-install components—more than the sum of its parts. Up to that time, factory power packs mainly consisted of multiple carburetors and a high-performance camshaft—usually with solid lifters, dual exhaust and sometimes high-rate valve springs. All of this was essentially simple bolt-on equipment you might buy in a hot-rod shop.

The '57 Corvette FI engine took a more-sophisticated approach. It wasn't only the fuel-injection system—though that was a very brilliant, efficient design. What made that Corvette power pack different was that the designers considered several *internal* components to improve performance *and reliability.* Internal parts included such things as domed pistons to bump compression from 9.5:1 to 10.5:1, lightweight valve-train components, heavy-duty aluminum main and rod bearings, and a deep, 5-qt oil pan.

When the complex Rochester fuel-injection system was added, you obviously had an engine that bore only superficial resemblance to the standard 283-CID passenger-car engine. The Corvette FI version had dozens of special parts not used in standard engines. In fact, the Corvette engine was so different that it had to be built on a special assembly line.

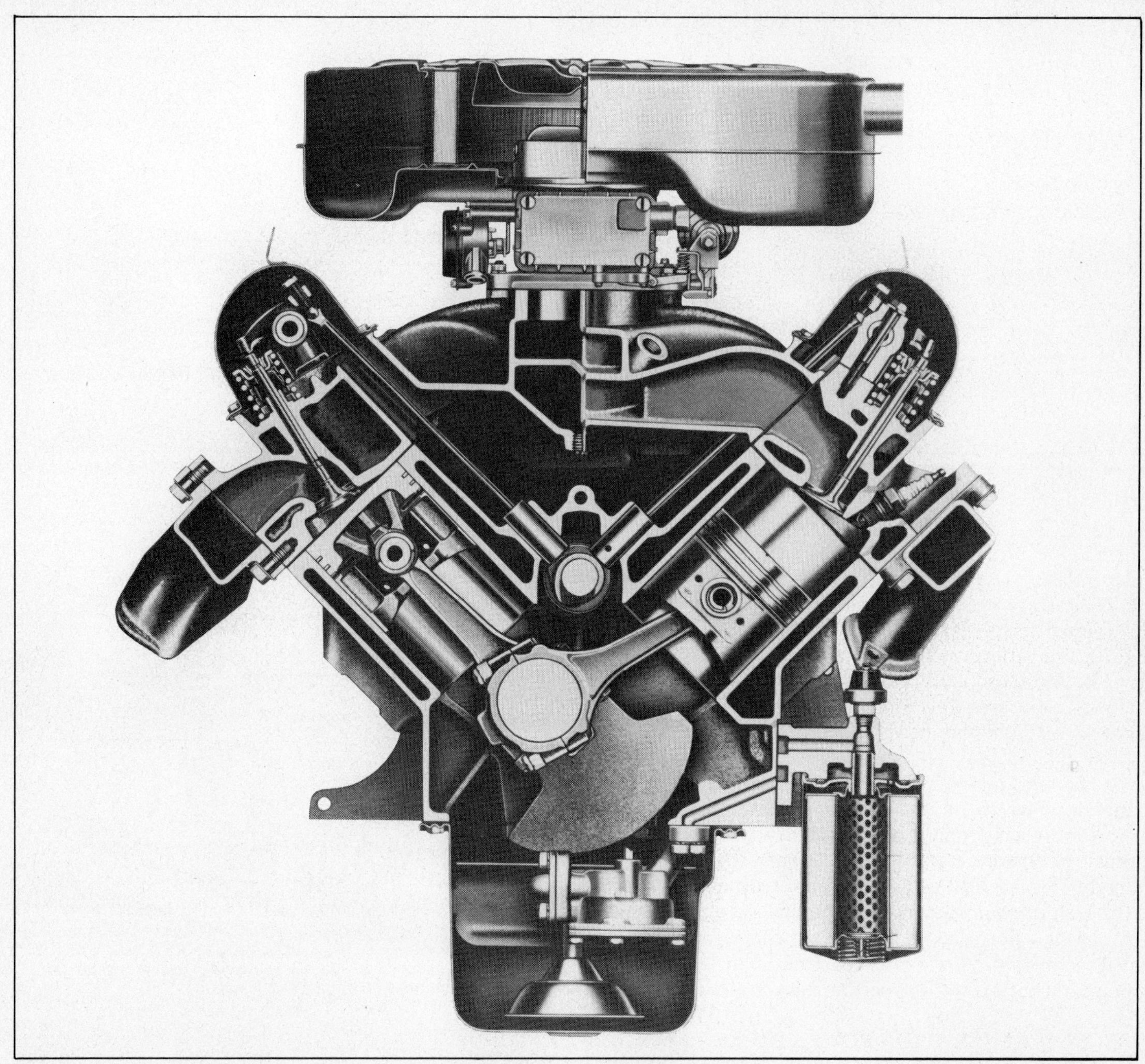

Some companies tooled up larger big-block V8s in the late '50s, as the trend to bigger displacements continued. This is Ford's original FE design. Introduced in 1958 at 332 CID, it eventually grew to 428 CID. Photo courtesy of Ford Motor Co.

This '57 Chevrolet is a good example of what I mean by a second-generation power pack: one that had a few more improvements than some bolt-on, hot-rod goodies. A second-generation power pack is one where consideration was given to high-rpm *durability,* not just high-rpm breathing.

When Chevrolet's first big-block engine, the 348-CID *W-block,* was introduced in 1958, the basic power pack was a three-2-barrel carburetion system, dual exhaust and nothing else. With hydraulic lifters limiting revs to 5200 rpm, you probably didn't need any additional bottom-end insurance. This engine was rated at a mild 280 HP at 4800 rpm.

But looking a little farther down the 1958 option list, there was a *Super Turbo Thrust* 348 engine, rated 315 HP at 5600 rpm. This is the second-generation power pack, based on the three 2-barrel-carburetion system. The package included 11:1 domed pistons, a special long-duration, solid-lifter cam, high-rate valve springs and aluminum rod and main bearings. The Super Turbo Thrust also featured a high-performance ignition system, with a centrifugal-advance distributor and high-output coil. The factory-approved red line was 6000 rpm. Well, at least you could rev it to 5500 rpm without holding your breath!

BIG BLOCKS TO THE FORE

It was probably inevitable that the horsepower race of the '50s would turn into a cubic-inch race. Traditionally,

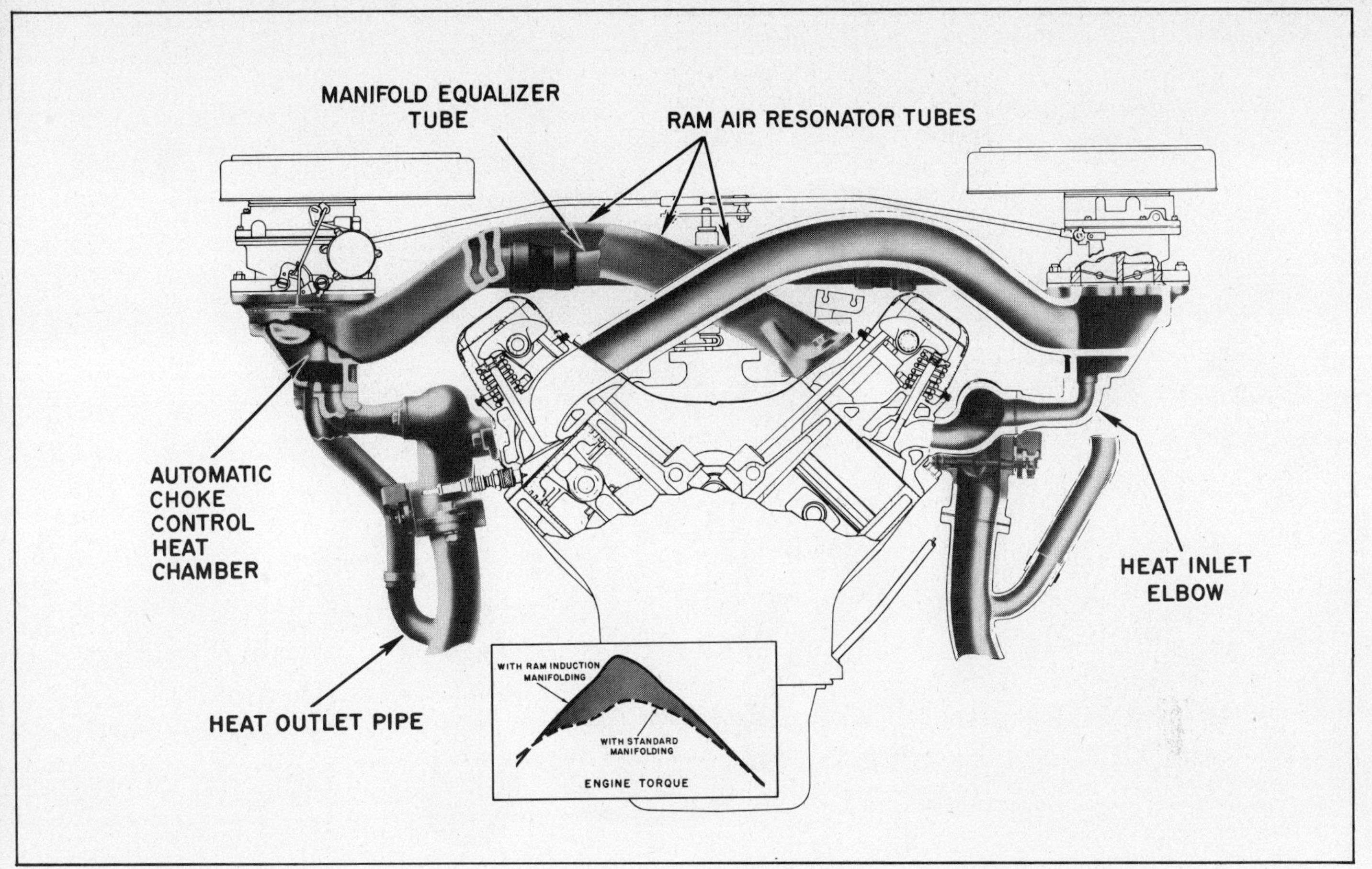

Chrysler engineers developed the concept of ram-induction manifolding in the late '50s. Long 30-in. manifold passages looped across the heads, giving a boost in mid-range torque from the bouncing pressure waves. Photo courtesy of Chrysler Corp.

American-car designers have followed the premise that a bigger bore and longer stroke were the easiest ways to improve performance—especially with cheap gasoline. New engines were always designed with an eye to expanding those dimensions.

But the trend to bigger cubes moved so fast in the mid-'50s that some companies began to retool for brand new big-block V8s within three or four years after their original overhead-valve V8s appeared. This would be unthinkable today. By the late '50s, practically all the Detroit companies were producing big-block V8s capable of future expansion to well over 400 cubic inches—even though their original displacements were closer to 350 cubes. The significant thing is that these engines had wide bore spacing and large crankcases to make room for future increases in bore and stroke.

Many Detroit engineers admitted they didn't know where displacement trends might go in the 1960s. They wanted to be ready for anything.

Needless to say, these large-displacement engines did a lot for street performance, and did so in cars weighing two tons and more. In these cars, street performance was as much a matter of mid-range torque as anything else. After all, torque is largely a function of displacement, so there's no substitute for cubes when you want quick tip-in response in a heavy car. You can feel the difference right away between, say, 300 and 400 CID—regardless of other factors like compression ratio, carburetion and cam timing. Then when you refine high-rpm breathing, you've got an unbeatable performance combination.

Thus the trend to big-block engines in the late '50s brought another significant increment in acceleration. A good example is that 348-CID Super Turbo Thrust Chevy mentioned earlier. I tested one in 1959 with 4.11:1 final drive and a three-speed manual transmission. The car clicked off 15.2-second quarter-miles all day at 93 mph, with closed exhaust. Despite the longer stroke, the engine would rev comfortably to 5800 rpm, thanks to good crank-train balance, high-rate valve springs and high-output ignition.

Small-block V8s were never quite the same again.

RAM INDUCTION

This next development didn't come until 1960, but it properly belongs in the power-pack section of performance evolution. I'm talking about the unique ram-type intake manifolds developed by Chrysler for some of their high-performance V8s.

Ram induction was a whole new theory of manifold design: Use equal-length passages for each cylinder, and then select that passage length to utilize the oscillating pressure waves on the intake stroke to *supercharge* the

Long-ram manifolding with dual 4-barrel carbs was adopted as standard equipment on Chrysler 300s of the early '60s. I think it was the most-impressive intake manifolding ever used by Detroit! Photo courtesy of Chrysler Corp.

cylinder by cramming in mixture just before the intake valve closes. These natural pressure waves travel at the speed of sound—about 1100 ft per second. Therefore, it's possible to calculate an optimum passage length that will get the reflected pressure wave from the primary suction wave, back to the cylinder just before the valve closes. But there's one catch. *You can ram-tune only for a specific rpm range.*

Admittedly, there were problems. To get the maximum ram boost at medium engine speeds, where it could help in passing acceleration on the highway, the optimum passage length figured out to about *30 in.* This would tune-in maximum ram induction between 2500 and 3000 rpm. Obviously, the specter of fitting 30-in.-long manifolds in the engine compartment presented no simple packaging problem. No wonder they called these early types *long-rams.*

Chrysler engineers finally came up with an arrangement using two 4-barrel carburetors. Manifold passages for one bank of cylinders looped across the opposite bank. The carburetor for each bank perched on a little plenum chamber on the outer end of the manifold casting, outside the opposite bank. Got that? As the drawing and photos show, there were separate castings for each bank. Each 30-in.-long manifold weighed 35 lb!

But what a tremendous torque wallop in the mid-range! The 413-CID wedge engines with these manifolds delivered nearly 500 ft-lb of torque at 2800 rpm. Long-ram engines were so responsive in the mid-range that you rarely had to downshift for highway passing. The standard procedure was to nail the throttle and hang on.

There was another problem with ram tuning at high rpm. The long manifold passages actually reduced breathing above 4000 rpm. To adapt the manifolds for high-revving engines, the castings were modified to remove part of the wall between adjacent passages. This reduced the *effective* passage length from 30 in. to 15 in., and pushed the maximum ram effect up to 5000—6000 rpm. These engines were called *short rams.* The ram effect was somewhat compromised with the extra turbulence in the common passages. But there was definitely a measurable boost in this higher-rpm range, adding at least 10—15 HP.

HOT ROD Magazine tested a '61 Chrysler with a 413-CID short-ram engine. Rated at 405 HP, the package included 11:1 pistons, 284° cam and dual exhaust. At the drag strip, it delivered a 14.73-second e.t. at 95.74 mph. The big, heavy Chrysler was fitted with high-traction tires and 4.56:1 final drive, so the combination was not strictly showroom stock. But it seems obvious from the strip times

Ramchargers racing club, a group of Chrysler engineers, first came into prominence in the late '50s and early '60s. This is driver Jim Thornton in their 413-CID '61 Dodge. Photo by Wally Chandler.

Long-ram manifolding on the 413-CID engine. Some racers chopped back the passage walls to shorten the effective ram length—to get maximum ram effect at 5600 rpm instead of 2800.

that the engine with factory exhaust system was delivering in excess of 300 net HP.

Chrysler engineers can take most of the credit for developing ram-type intake manifolding for the American auto industry.

AT THE END OF THE DECADE . . .

At the end of the 1950s, performance development in Detroit was mostly a backroom activity. The AMA anti-racing resolution of June, 1957 was still very much in effect. None of the auto companies were openly engaging in speed events or running any wild, performance-oriented advertising. On the surface, at least, the companies appeared to be de-emphasizing performance and horsepower.

But plenty of development was going on behind the scenes. NASCAR racing was still flourishing. And most of the large-volume car divisions were interested in making a good showing. Ford and Chevrolet were the top winners with their new big-block engines, with Dodge and Pontiac right behind.

But development of this NASCAR equipment had reverted to the situation that existed in the early '50s. That is, the parts were available over-the-counter only—not as optional assembly-line equipment. About the only difference is the carmakers called it *export* equipment in the early '50s, and *police* equipment in the late '50s. The ultimate purpose was still the same: improve all-out racing performance on the NASCAR tracks.

I mentioned earlier that NASCAR rules were changed in 1957 to limit all Grand National cars to one 4-barrel carburetor. This rule had a substantial impact on optional performance equipment available for the street. That is, without the incentive to develop multiple-carburetor systems for NASCAR racing, some of the companies backed off on factory-installed power packs. This was the case with Ford and Oldsmobile. But Chevrolet, Pontiac and Chrysler continued with multiple-carburetor systems and power packs—even the sophisticated second-generation power packs mentioned earlier. Chevrolet, Pontiac and Chrysler had two parallel performance programs: over-the-counter NASCAR goodies plus the optional factory-installed power packs. Ford and Olds concentrated on the special NASCAR stuff.

If I had to pick a *street-performance* leader at the end of the '50s, it would be Chevrolet. Not only did Chevrolet offer sophisticated power packs for the 283-CID small-block V8, but by 1959, they had done just as much development on the 348-CID W-block. As mentioned earlier, available big-block equipment combinations could give quarter-mile times in the mid-15s at over 90 mph in late-'50s full-size Chevys. Small-block power packs were a second quicker and 6—8-mph faster in the lightweight Corvette.

I've called these high-performance engine options *power packs*. You can call them *police engines, super-duty engines* or whatever you want. The point is that you could order them in your new 1959 Chevrolet for less than $150 extra. You'd have enough instant performance to out-accelerate and out-top-speed almost anything on the road with simple, factory-developed, optional equipment.

That was the power-pack era of Detroit's performance history.

Action at the 1966 NHRA Nationals pits a '63 Plymouth Super Stock against a '66 Plymouth Belvedere. Photo by Wally Chandler.

3

Factory Super Stocks

Engineered for performance

FORD BREAKS THE ICE

It's ironic that the company that most faithfully adhered to the AMA anti-racing resolution of 1957 was the first to bust it wide open in 1960. But automotive historians generally agree that it was Ford's impressive 352-CID high-performance option package of 1960 that broke the ice for more-open promotion of performance and horsepower in the marketplace. Historians also agree that this Ford engine paved the way for increased factory involvement in NASCAR and drag racing.

In the preceding chapter, I mentioned that during the late '50s, most auto companies did their performance development behind the scenes. Their most-potent performance parts were only available over the counter: You installed them yourself. Ford was so conscientious about this that they didn't even offer an exciting power pack on an assembly-line model. Actually, Ford had only recently joined the AMA. The elder Henry Ford had snubbed the industry organization for years, but apparently Henry II felt obligated to follow its rulings to the letter.

But Chevrolet's quick takeover of the youth market after 1955—plus its swift buildup of a solid performance image—was too much for the younger Ford to swallow. This is the image that Ford owned in the '30s and '40s. Now, Chevrolet had stolen it away from them in four short years with new V8 engines that were just as smooth, quiet, economical and practical as anything Ford offered. But cubic-inch for cubic-inch, Chevrolet street engines offered substantially better performance. Henry Ford II was not going to take this without a fight, AMA or no.

Above: Ford made the first true super-stock engine from their 352-CID FE block in 1960. Dubbed the *352 Special*, it was re-engineered from the ground up for race performance. Note beautiful exhaust manifolds. Photo courtesy of Ford Motor Co.

Shortly after the new big-block *FE* engine was introduced in 1958, Ford turned his engineers loose on a *real* high-performance street version of the 352-CID engine.

FROM POWER PACK TO SUPER STOCK

I'll call this new type of car and engine a *super stock*. Super-stock engines were even more sophisticated than the second-generation power packs of the late '50s. They were standard big-block V8s literally re-engineered for performance.

The new super-stock engines had more special pieces—not only bolt-on stuff. The designers went through the engine, checking each part for stresses under sustained full-throttle operation at the factory-designated rpm limit. When standard parts broke, they were replaced by special heavy-duty pieces.

I'd be lying if I said that first Ford super-stock engine of 1960 was a good example of this new performance concept. I guess the only way we can excuse it is to say that the Ford engineers had been away from street performance too long.

The cylinder block and pistons were standard 352 fare—with only a stiffer oil-pump-relief spring to raise oil pressure from 55 to 65 psi. But the rods were heavy duty, with a wider beam. Late versions had larger rod bolts—13/32 in. vs. 3/8 in. And the crank had larger counterweights to go with the heavy-duty rods. A factory oil cooler was available with the engine, too.

Cylinder heads were similar to stock heads, but with larger ports and smaller combustion chambers to raise compression ratio to 10.6:1. Ford used a high-performance, 306°, solid-lifter camshaft with 0.480-in. valve lift—but then closed the valves with standard springs! The valves would float at 5200 rpm. At least the weak valve springs kept you from breaking anything in the bottom end!

On the outside, the high-performance 352 Special had lots of eye-appeal—right down to the aluminized valve covers with big decals. The engine featured a special, big-passage aluminum intake manifold. On it was mounted a neat 550-cfm Holley 4-barrel. Above this was a broad, open-element, paper air filter with a chromed top.

The exhaust system was worth looking at, too: Huge, cast-iron low-restriction manifolds on each side dumped into large 2-1/4-in. headpipes and low-restriction mufflers. This exhaust system was a breakthrough in the sense that it used special exhaust manifolds. All previous factory-installed power packs had used standard production manifolds. These new Ford castings were things of beauty—though they added nearly 50 lb.

Ford officials advertised their new 352-CID Special at 360 HP at 6000 rpm. They wanted to upstage Chevrolet's power pack for the 348-CID W-block—rated 350 HP at 5600 rpm. But Ford couldn't back it up on the drag strips. Early models of the 352 Special were a disappointment because of the low-rate—190 lb at full lift—valve springs and the resulting limited rev range. Within weeks of the engine's debut, Ford engineers released higher-rate springs that gave a 270-lb force at full valve lift. The new springs permitted up to 5800 rpm with solid lifters. With the engine now able to rev higher, acceleration with the three-speed manual transmission was greatly improved.

I had a chance to test a '60 Ford two-door with this engine, using street tires and dealer-installed 5.14:1 final drive. I had to feather the clutch on starting to keep from smoking the tires, but got these times when shifting at 5600 rpm:

0—60 mph	8.2 sec
1/4 mile e.t.	15.9 sec @ 89 mph

I also took some readings with an accelerometer in 2nd gear, which permitted calculating the actual horsepower output—*net horsepower*—at the flywheel, as installed in the car. See the accompanying sidebar for further explanation of net horsepower. For the Ford 352 Special engine, this figured out to about 270 HP at 5400 rpm, with factory exhaust and air cleaner.

It's obvious from these performance figures that the new Ford 352 Special was no stronger than the simpler, second-generation Chevrolet 348 power pack of the late '50s. That Chevy could also turn quarter-miles down in the 15-second bracket with terminal speeds around 90 mph. And its net horsepower per cubic inch was perhaps a bit better than the Ford.

WHAT ABOUT NET HORSEPOWER?

What about those *net*-horsepower figures sprinkled throughout the book? We know that net HP means the actual power *delivered at the flywheel,* as installed in the car, with all standard accessories and exhaust system in place. But we also know such figures were never published by the car companies until the early 1970s and here are net figures for engines in the '50s and '60s. How come?

These were calculated on my own private "dream wheel." But it's all scientific. Actually, I tested dozens of high-performance cars for true net HP during the supercar period. It can be done with an instrument called an *accelerometer.* It measures the instantaneous rate of acceleration or deceleration. If you take a reading under full throttle at some known speed—then another reading at that same speed when coasting—the total developed horsepower can be calculated from two simple equations:

$$\text{Force (lb)} = \text{Acceleration (g's)} \times \text{Weight (lb)}$$
$$= \frac{\text{Acceleration ft/sec}^2}{32.2\ \text{ft/sec}^2} \times \text{Weight (lb)}$$

$$\text{Power (horsepower)} = \text{Speed (mph)} \times \text{Force (lb)} \div 375$$

There's a little more to it than this. You have to make reasonable allowance for rotating inertia of the drive line, and drive-line friction. But with a little experience, test results should be close to published net horsepower figures by using this method.

But there's more. I didn't test all the cars in this book. The beauty of the net-horsepower figure is that there is a very close correlation between the pounds-per-horsepower ratio for a car and the terminal speed of that car on a quarter-mile drag strip. In other words, when I plotted the pounds-per-horsepower figures against quarter-mile terminal speeds, the points fell very close to a single curved line. I could even derive a simple mathematical equation for that line.

Now you see the trick: Because I knew the terminal speeds and weights of all the cars here, I merely had to read off the pounds-per-HP ratio from the graph. Then divide that into the weight—*and you've got the true net horsepower.* Simple—but it took 20 years of testing to get that graph!

But this 352-CID Ford was still a milestone engine in that it used special, low-restriction exhaust manifolds. In this sense, it started a trend that was to grow and flourish for another 10 or 12 years in the Motor City. The 352 Special Ford started the modern supercar era of American automotive history. Even though this first example was admittedly not all that sensational, it was a beginning.

TWO TYPES OF PERFORMANCE ENGINES

After Ford set the pattern for the fully engineered production performance engine in 1960, engine development seemed to split into two distinct paths. That is, there were engines developed strictly for the *street,* and engines strictly for the *race track.*

This concept of a factory-produced full-race engine was also brand new at that time. Up to then, factory equipment

Bill Jenkins' "Old Reliable" 409 Chevy with Dave Strickler at the wheel, in a go with Sidwell's Pontiac at Detroit Dragway. The Jenkins/Strickler team was dynamite in those days. Photo by Wally Chandler.

designed strictly for racing was sold over-the-counter, to be installed and tuned by the private mechanic. All of a sudden, here was this idea of assembling highly tuned, all-out race engines on factory production lines, installing them in selected models and selling them to the public out of dealer showrooms. Admittedly production rates were very low on these engines, car prices were high and only a few selected buyers were given the opportunity to order. But the fact remains that there was a time in Detroit history when some of the car companies built a few-hundred all-out racing cars on their assembly lines!

This sounds so incredible today that you have to think about it a minute. What market and economic conditions could have possibly led the car companies to assemble all-out racing cars? The special tooling and equipment alone must have lost many thousands of dollars for the companies in terms of unit price vs. cost! It's especially curious because many professional racers would have been glad to buy the parts over-the-counter and assemble the cars themselves—at much-less cost to the factory.

I think there were three reasons for this unlikely situation: (1) Factory-built race cars enhanced the company's performance image in the youth market; (2) It got a few more of these cars on the race tracks more quickly than if privately built; and (3) It permitted these cars and engines to qualify as *stock* in the eyes of the race-sanctioning organizations. This last factor was very important in drag racing, where there were many classes for modified cars, and the definition of a stock car was narrow.

Whether these reasons were strong enough to justify the thousands spent on factory racing cars, only history can judge. But it leaves no doubt about the strength of the youth/performance market at that time. This segment of the overall U.S. car market was absolutely boiling in the late '50s and early '60s. And it was a *growing* segment because of earlier birth-rate patterns. Millions in profits were waiting for the company that could appeal to young people in a big way.

It's a fact that Pontiac jumped from 6th to 3rd place in sales in only three years, primarily by building cars that appealed to young drivers. Pontiac's meteoric rise didn't go unnoticed by the other companies. They all wanted a piece of the action. By the early '60s, the fight was joined.

Is this enough reason for factory-built race cars?

With the optional 409-CID V8, the full-size '61 Impala enjoyed supercar acceleration despite its 3700-lb weight. Photo courtesy of Chevrolet.

Original 409 Chevy engine of 1961-1/2 used a single 600-cfm Carter AFB carb on a new aluminum manifold. Otherwise it was similar to the Super Turbo Thrust 348s of the late '50s. The 409 could turn 6200 rpm with solid lifters. Photo courtesy of Chevrolet.

Chevy 409 engine had heavy-duty rods and crank, high-output oil pump, and was the first production engine to use a windage tray under the crankshaft. Suitable for both street and NASCAR tracks. Photo courtesy of General Motors.

STREET OR RACE TRACK?

With this new concept of the factory-built race car, companies were faced with the choice of where to spend their performance-development budgets to get the best impact on sales and profits: For street performance? For a race-track image? Or maybe both?

As it turned out, Chevrolet and Ford stuck pretty much to the street in the early '60s. They did develop plenty of race-track hardware, and even built some complete race cars in the '62—'65 period. But in the early '60s, they put their money and image on the street. The race goodies were mostly over-the-counter.

Chrysler and Pontiac took a different tack. For street fans, they offered what I call *semi-performance* engines. These were engines with reasonable compression and hydraulic-lifter cams, but with impressive carburetion and image-boosting external goodies.

These companies did spend a lot more money on their race cars. In the'62—'64 period, Chrysler and Pontiac released complete big-block packages that were highly competitive both on the drag strips and NASCAR tracks. And just to hedge their bets, they cleverly made these race cars to be driven legally on the street—if you wanted to put up with a rough idle, poor throttle response, noise, vibration and terrible gas mileage. This was done by offering optional 11:1 pistons, instead of the usual 12.5:1 race pistons, and by installing mufflers and capped exhaust cut-outs under the car. Perhaps overly optimistic company officials imagined these cars could be world-beaters on both the street and the race tracks!

Chevy 409s of the '62—'63 period had two Carter AFBs and new cylinder heads with larger ports and valves. They were at least 20- or 30-HP stronger than earlier single-carb jobs. Photo courtesy of Chevrolet.

It's hard to believe this ever happened in Detroit. But it did—and not too many years ago.

ENGINES FOR THE STREET

Let's take a more-detailed look at the street engines offered by the various companies in the early '60s.

Chevrolet—The 409-CID Chevrolet W-block of 1961-1/2—the fabled 409 Chevy of song and verse—was one of the best-known high-performance street engines to

Early 409 Chevys were very successful on the drag strips. That was the day when racers lowered the rear suspension, thinking the resulting "cowboy rake" gave better traction. It didn't. Photo by Wally Chandler.

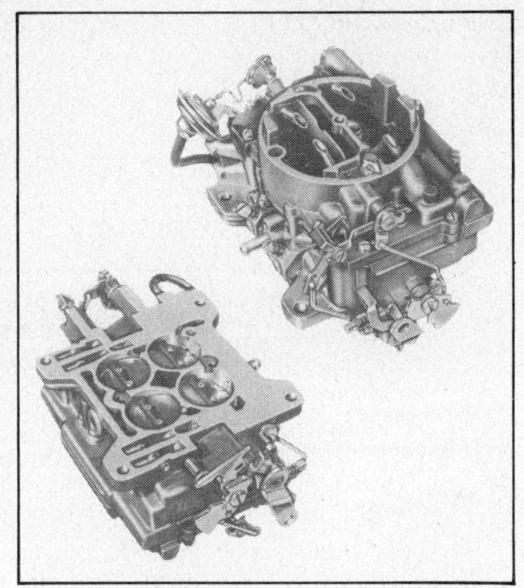
Carter's *AFB*—aluminum four-barrel—was introduced in 1961. The AFB was a key factor in the good performance of many super stocks of the early '60s. Its 600-cfm flow-capacity rating was higher than any other 4-barrel at that time.

come out of the supercar era. The original version was essentially a bored-and-stroked 348, but with the bottom end beefed up with stronger rods, harder bearings, forged pistons, 6-qt oil pan and a *windage tray* under the crankshaft. The tray reduced oil aeration and increased power output by reducing the amount of oil being whipped by the crank. It reduces the amount of oil sticking to the crank like caramel to an apple. It was one of the first uses of this clever device in a production engine.

Up on top, the 409 featured 11:1 compression, large-passage aluminum manifold, the newly introduced 600-cfm Carter AFB—aluminum 4-barrel—carburetor, and special low-restriction exhaust manifolds. A long-duration solid-lifter cam was used with standard valves and high-rate springs. As with the 348, ignition was by a dual-breaker, centrifugal-advance distributor. Red line was set at 6000—6200 rpm.

Important improvements were made to the 409 in '62 and '63. These included new cylinder-head castings with larger ports and valves, dual 4-barrel carburetion, redesigned piston domes for better combustion, a redesigned camshaft, and still-higher-rate valve springs with dampers. This pushed the red line to 6500 rpm. Oil pressure was raised to 75 psi. Interestingly enough, Chevy reverted to a single-breaker, vacuum-advance distributor for '62 to improve around-town fuel economy and response. Nevertheless, these last 409s were at least 20—30-HP stronger than the early single-4-barrel version.

Dodge/Plymouth—As previously mentioned, Chrysler's big gimmick for high-performance street engines in '60—'61 was the ram-type intake manifold with dual 4-barrel carbs. This impressive induction system was offered on several big-block V8s. A choice of manifolds was offered with either long or short ram lengths to give ram boost at medium or high rpm, respectively. These were coupled with high-compression pistons, a high-performance hydraulic cam, and a low-restriction, dual exhaust system. Some, like the short-ram, 405-HP, 413-CID com-

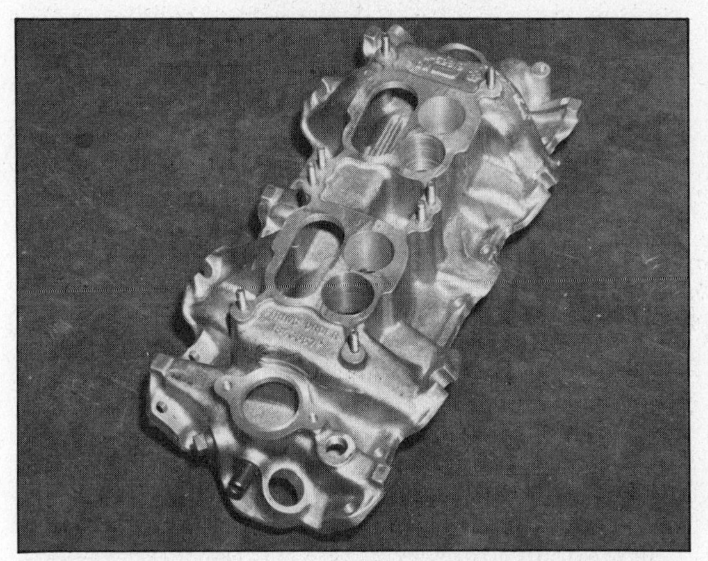
Dual-4-barrel manifold used on 409 Chevys had all left-side barrels feeding the lower plane of passages, with right barrels feeding the upper plane—rather than usual side-by-side arrangement. It worked better than other manifold systems with the Carter AFBs. Photo courtesy of Chevrolet.

Dodge and Plymouth offered street versions of the 426-CID wedge engine in '64 and '65. This is a '64 Plymouth Sport Fury convertible with the 426-S option. Photo by Ralph Ronzello.

Detroit-area Ford salesman Jack Gray publicized his products with an active drag-racing program in the early '60s. High-performance cars were just beginning to make money in those days. The best was yet to come. Photo by Wally Chandler.

Cutaway of Ford 406 Special of 1962—'63. Single 4-barrel version was rated at 385 HP. Three-2-barrel version was advertised at 405 HP. Photo courtesy of Ford Motor Co.

bination, were fairly strong engines. But they never gained the reputation or following the Ford and Chevy street engines had during that period.

Ford/Mercury—I've described the 352-CID high-performance engine Ford introduced in 1960 as the first of a new breed of fully engineered super-stock street engines. Although the 352 Special was not a great engine, improvements in its basic design were carried over and expanded upon in future versions of the Ford FE engine. These design changes helped put Ford right in the thick of the performance race during the next two or three years.

In 1961, displacement was expanded to 390 cubic inches. The bottom end was completely re-engineered to improve durability at high revs, both for drag racing and NASCAR. A new block casting had thicker main webs, ribbed walls and larger oil passages. Main-bearing journals were grooved to increase oil flow. Connecting rods were reinforced and fitted with larger bolts. The engine was said to be safe to 6200 rpm, despite the cast-iron crankshaft.

In 1962, Ford introduced the famous 406-CID HP engine. The extra cubes, plus larger exhaust valves and a beautiful new triple 2-barrel carburetion system, boosted output at least 20—30 HP over the 390. The three Holley 2-barrel carbs had a total airflow capacity of 920 cfm. They were mounted on a big-passage, aluminum two-plane manifold. The carbs worked through a progressive throttle linkage to give flexible response and good fuel economy on the street.

Toward the end of the '63 model run, Ford released a street version of their 427-CID NASCAR engine. The big feature was a new dual 4-barrel carburetion system, using two 600-cfm Holleys with progressive primary throttles and vacuum-operated secondaries. It was undoubtedly the smoothest, most-flexible high-volume carburetion system to come out of Detroit up to that time.

Here's how it worked: At very small throttle travel, only the primary butterflies in the front carb were open. When these reached about one-third open, the primaries in the

Triple carburetion system for the 406 Ford used three Holley 2-barrels on an excellent two-plane aluminum manifold, with mechanical throttle linkage. Total flow capacity was about 920 cfm.

Ford dressed up their 406 4-barrel engines with aluminized valve covers—the three-2-barrel jobs got gold valve covers. Both 406s had a beautiful cast-aluminum air-filter housing. Detroit manufacturers were very conscious of the image their engines projected.

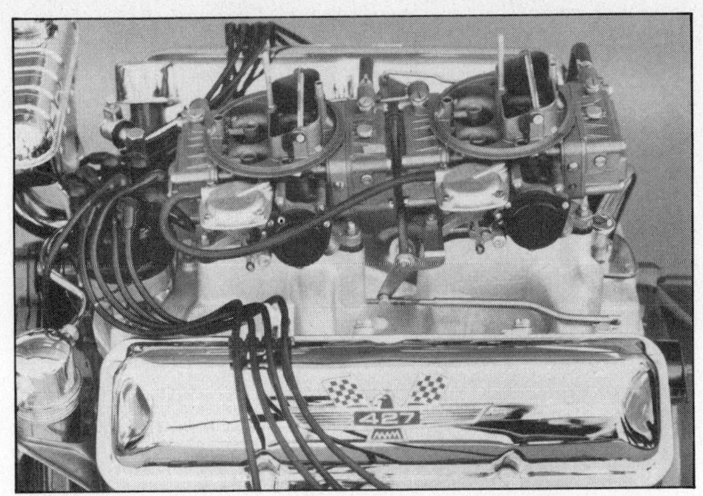

Ford used two 600-cfm Holley 4-barrels to fill the cylinders on the big 427-CID V8 of the '63-1/2—'67 period. Note that the secondary barrels are controlled by vacuum diaphragms.

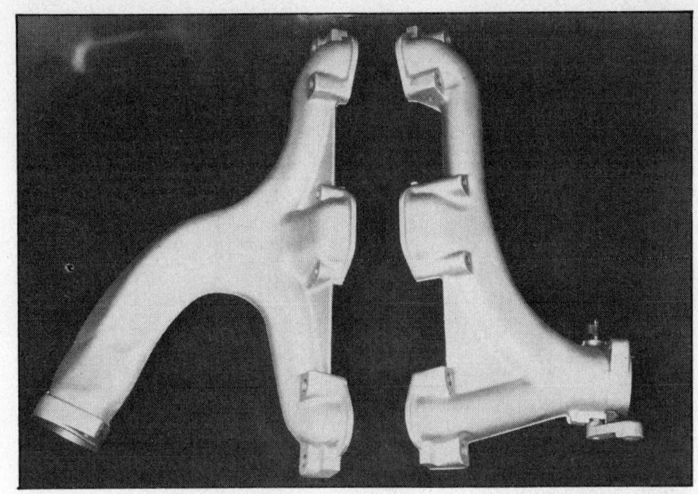
Pontiac used these beautiful low-restriction exhaust manifolds on the 389-CID Trophy engines in the early '60s. The factory manifolds were an important performance factor, as bolt-on headers were not as popular in those days.

Pontiac's strongest street combination in the early '60s was the 389-CID *425 Trophy-A* engine with Tri-Power carburetion and hydraulic cam. A full-size Catalina coupe was the lightest model you could get it in. Remember, this preceeded the GTO. Photo by Wally Chandler.

Optional *Tri-Power* carburetion system was very popular on 389 Pontiacs in the early '60s. Most owners converted from electro-vacuum to mechanical throttle linkage. Photo by Alex Walordy.

rear carb started to open—but at a faster rate so all four primaries reached wide-open throttle together. Meanwhile, the secondary butterflies in both carbs were controlled automatically by diaphragms that sensed the vacuum in the primary venturis. Primary-venturi vacuum increased as airflow increased. Thus, the secondary throttles would open gradually between about 2000 and 4000 rpm as engine speed increased. They fed extra air and fuel only as the engine could *use* it, depending on the calibration of the vacuum diaphragms. A neat system. Holley engineers had used the system on single 4-barrel carbs for several years. It seemed to work just as well with duals.

Other than this unique carburetion system, there wasn't much difference between the 406 and 427 Ford high-performance engines. Early street 427s used 406 heads, camshaft, exhaust system, and so on. But in '65—'67, several goodies from NASCAR engines were adopted on the street versions: larger ports and valves, higher-lift and longer-duration cam, forged-steel crank and forged pistons. Also adopted were cross-bolted mains developed for NASCAR 406s in late '62. These were some of the strongest, most-rugged street engines of the supercar era.

This Ford FE-engine high-performance program was one that evolved over a long period of time—say seven or eight years. It wasn't one that bloomed quickly and flickered out, like the 409 Chevy. Late examples of the FE engine were truly sophisticated high-performance street engines.

Pontiac—Pontiac's premier street-performance engine in the early '60s was based on the 389-CID block. It was known as the *Trophy 425-A* engine. The *425* stood for the engine's torque rating, not horsepower or displacement. Compared with the 409 Chevys and 406 Fords, the 425-A Pontiac was a fairly mild engine. It used standard cylinder-head castings, 10.25:1 compression, a medium-lift hydraulic cam, and the triple 2-barrel carburetion system that had

Car	Transmission	Axle ratio	Adv. HP	0—30 mph	0—60 mph	1/4-mile e.t. @ mph	Est. Net HP
'61 Chevrolet 409	M4	3.70:1	380	3.0 sec	7.3 sec	14.9 @ 94	290
'61 Dodge 413	M3	4.10:1	385	3.5	7.4	15.1 @ 92	270
'62 Ford 406	M4	3.56:1	405	2.9	7.0	15.3 @ 93	300
'62 Pontiac 389 (tuned)	M4	3.90:1	348	2.5	6.6	14.5 @ 95	320
'63 Mercury 427	M4	4.11:1	425	2.7	7.0	15.1 @ 87	270

been developed in the late '50s. It did, however, sport a reinforced cylinder block and four-bolt main-bearing caps. This block was a carryover from the NASCAR program. Pontiac engineers also designed special low-restriction exhaust manifolds for the Trophy 425-A engine—which set it apart from the bread-and-butter Pontiac engines in the early '60s.

The throttle linkage for the 6-barrel carburetion system warrants special mention: It was probably as bad as the 427-Ford linkage was good. The Pontiac accelerator-pedal linkage operated only the center 2-barrel carb. As long as wide-open throttle was avoided, the engine breathed through the center carb—even up to 80 or 90 mph. But if enough throttle was applied to open the center carburetor's butterflies about three-quarters, the linkage tripped a vacuum switch. This switch vented engine vacuum to a diaphragm on the two end carbs, and flopped their butterflies wide open *right now*. Got that? In one big WHOOOSH you had another 20—30 horses pulling. The power came in with a punch that would press you back in the seat and scare the pants off unsuspecting passengers.

The Pontiac 389-CID V8 with Tri-Power was a fun plaything for the wild-eyed street crowd. But serious street and drag racers didn't like the setup because you couldn't control the torque precisely for fast shifting and feathering off the line. It was either all or nothing. Racers would invariably replace the vacuum linkage with a mechanical setup—which, significantly, was offered over-the-counter with a Pontiac part number. The designers knew the vacuum setup was no good, too!

MEASURING 'EM UP

The real measure of any high-performance street engine is the acceleration it provides in a standard body, with typical street tires and gearing *as it comes off the showroom floor*. This is the way the vast majority of supercar fans used their trusty steeds in the '60s. The professionally tuned car with special equipment and drag-strip gearing was the exception.

For comparative performance figures on these cars, I'll rely heavily on published road tests made by CAR LIFE magazine in the 1960s. CAR LIFE had a flawless reputation for honest, unbiased road tests on Detroit models in those days. Furthermore, CAR LIFE invariably tested production models, with regular factory equipment and no special tuning.

Many of the other car-buff magazines would test models that were professionally tuned, with special gears and slick-tread tires—specially designed for high traction. Sometimes, magazines would test them with the factory exhaust manifolds replaced by open, tubular-steel headers. When tests were conducted in such a loose manner, the figures didn't tell you how the average car would perform on the street.

Above are CAR LIFE test figures on some of these early-'60s super stocks.

Net horsepower is estimated from terminal speed in the 1/4 mile, based on the total car weight.

You can see from these figures that those early-'60s super stocks would average 0—60-mph times in the 7-second bracket. Quarter-mile elapsed times in the low 15-second bracket at over 90-mph trap speed were common. Net horsepower outputs were in the general range of 270—300 HP.

Don't take too much stock in the times shown for the '62 Pontiac 389. The test car had the famous *Bobcat* tuneup package from Royal Pontiac in Royal Oak, Michigan. In the early '60s, Royal was the largest seller of high-performance Pontiacs in the country. Royal-modified Pontiacs were often favorites of the infamous Detroit-area Woodward Avenue street-racing crowd.

The Bobcat kit included a modified spark-advance curve, richer carburetor jetting, blocked manifold heat, re-indexed cam timing—one-tooth retarded so valves would open and close later to move the power curve up in the rev range—and the aforementioned mechanical throttle linkage. The Bobcat treatment could be had for an extra $175. It gave a Trophy 425-A engine a little edge over the other Ponchos on the street.

Royal Pontiac also performed what they called a *lifter-travel adjustment*. They backed off the rocker-arm adjustment to bring the hydraulic-lifter plunger to the top of its travel, up against its retaining ring. The adjustment gave the effect of solid lifters and was said to be good for an additional 300 rpm on the top end.

Though most of these changes hurt low-speed performance, power was up noticeably in the mid-range, all the way up to 5700 rpm. The changes improved high-speed response and shifting. They also added an apparent 30—40 HP over a showroom Trophy engine. The CAR LIFE test car turned 0—60 and 1/4-mile times more than a half-second quicker than a typical factory model.

Likewise, the '63 Mercury 427 shouldn't be considered a representative model. The test car was a two-ton convertible and the engine didn't feel strong to the testers. Perhaps it suffered from some minor tuning foul-up. The 4.11:1 final drive helped salvage some performance, but that net horsepower should have been closer to 300. In fact, the output of the '62 Ford 406 should probably have been *less* than 300 HP—or at least less than the '63 427 engine. These factory supercars varied a lot in performance, like factory cars do today. It's the old story of trying to buy a car built on a Tuesday, Wednesday or Thursday—they're more-carefully assembled than Monday- or Friday-built cars.

THOSE FACTORY RACING CARS

Even though the factory-built super-stock racing cars mentioned earlier were never intended for street driving, they had a vital influence on the super-performance cars that *were* engineered for everyday driving. And as far as that goes, these factory race cars were offered with compression ratios *reduced* to around 11:1, and with street-legal mufflers hooked up. More than a few well-heeled buyers with connections bought them primarily for serious street racing. These streetable super stocks were definitely part of the grass-roots supercar scene in the early '60s, whether confined to the race track or in a quick midnight bash on a blocked-off freeway.

Chevrolet, Chrysler, Ford and Pontiac turned out these special limited-production race cars in the early '60s. They were essentially highly modified production engines and heavy-duty drive lines installed in the lightest two-door bodies. Heavy-duty suspension components were thrown in to help get the power to the ground. But the packages were fairly complete. Except for installing slicks and uncapping the exhausts, the cars were supposed to be ready to race on the drag strip. I should note, though, that many minor modifications were allowed on stock cars by the various race organizations. Top competitors used every trick the rules allowed.

The major drag-racing sanctioning body at that time, the NHRA, had no minimum-production requirements for factory-stock models. Anything that came off a factory line would fall into one of the stock classes, based on advertised horsepower and shipping weight. Thus, the factories could get away with producing only a handful of cars—maybe only 30 or 40.

Later on, NHRA set minimum-production limits: first at 50 units; and later at 100. But these small numbers didn't give the factories any real production or distribution headaches. The performance-image rub-off was considered well worth the cost and effort to hand-assemble these special cars.

Consider some of the packages from the various companies . . .

Pontiac—I'm listing Pontiac first because they started this trend to the race-only super stock. Their first effort in 1962 was probably an answer to the 409-Chevy street jobs—but with more of everything. Essentially, what they did was install over-the-counter NASCAR parts in the 389-CID engine, bore and stroke it to 421 CID, and put it in special lightweight Catalina two-doors.

The engine package included dual 4-barrel carburetors on an unheated aluminum manifold, big-port NASCAR heads, 12.5:1 compression, forged pistons, rods and crankshaft, new camshaft and dual valve springs, and a new set of *4-into-2* split-flow, dual-outlet exhaust manifolds. The drive line consisted of a heavy-duty clutch, manual four-

Jim Wangers won Super Stock Eliminator at the NHRA Nationals in 1960 with a Super Duty 389 Catalina, turning low 14s at 102 mph. Drag racers used all factory equipment in the stock classes in those days. Photo by Wally Chandler.

Jim Wangers, left, and crew chief Frank Rediker with Royal's lightweight '63 Catalina Super Duty 421. Wangers won the B/FX class at the NHRA Nationals with the car that year. Photo by Wally Chandler.

Pontiac's Super Duty 421-CID engine of the '62—'63 period used two Carter AFBs and big-port cylinder heads. It developed more than 450 HP with factory manifolds; would rev to about 6200 rpm. Photo courtesy of Pontiac.

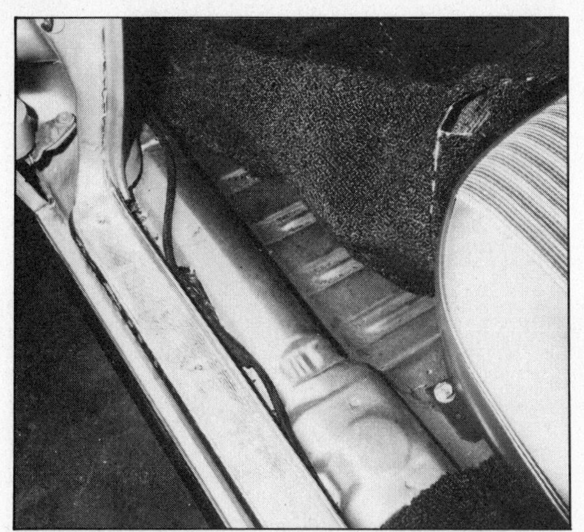

Super Duty 421s of 1963 saved 100 lb by omitting all body sealer, sound deadener and insulation. They rumbled like a bass drum going down the street! Photo by Wally Chandler.

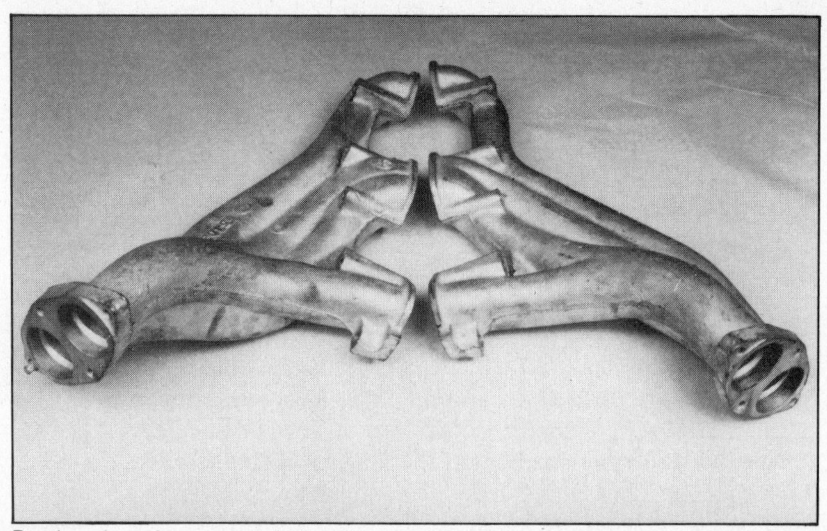

Pontiac plowed new ground when they cast the Super Duty 421 exhaust manifolds out of aluminum. The manifolds saved 45 lb—but left a few drops of aluminum on drag strips every time they ran! Pontiac went back to cast iron the following year. Photo by Wally Chandler.

Moving the battery back to the trunk was a favorite way to improve weight distribution in the early '60s. They did it at the factory on Super Duty 421 Pontiacs. Photo by Wally Chandler.

In '63, Pontiac went a step beyond with the factory exhaust system by providing a *cutout* on the ends of the manifolds. Covering each cutout was a plate that could be removed to bypass the mufflers. Photo by Wally Chandler.

speed transmission, oversize drive shaft and U-joints, and special heat-treated ring-and-pinion gears with a limited-slip differential.

The body package deserves special mention. That is, full-size '62 Pontiacs were relatively heavy cars—around 3800 lb for a stripped Catalina coupe. Pontiac engineers knew this was too heavy to be competitive with the other super stocks of the day.

They had recently been experimenting with techniques for stamping sheet aluminum on production body dies. So the idea was a natural: Stamp out a batch of special aluminum body panels for the new 421-CID super stocks. Eventually, they used aluminum hoods, inner and outer front-fender panels, front bumpers and radiator brackets. In total, the engineers eliminated 200 lb. Also, weight distribution was improved by taking the bulk of this 200-lb weight reduction off the front wheels. The big '62 Ponchos were much more competitive as a result.

The 1963 package was many times more sophisticated. The engine was improved with bigger carburetors, bigger valves, a deep oil pan and high-output oil pump. Pontiac's resident cam man, Malcolm McKellar, added a new camshaft and dual valve springs that extended usable revs from 6000 to 6300 rpm. These modifications added at least 20—30 HP with open exhaust.

But the big story in 1963 was in the weight-reduction program. Chief Engineer John DeLorean and his cohorts got completely carried away. In addition to the '62 aluminum body panels, Pontiac used a cast-aluminum bellhousing, transmission case and tailshaft housing, and rear-axle center section. Another 45 lb was eliminated by casting the huge exhaust manifolds out of . . . hang on to your hat . . . aluminum! By the way, these cast-aluminum exhaust manifolds had a usage restriction—for drag-strip runs not to exceed 14 seconds, lest they start to melt and drip on the paving!

The weight-reduction program continued with the chassis. The frame was drilled so full of lightening holes it

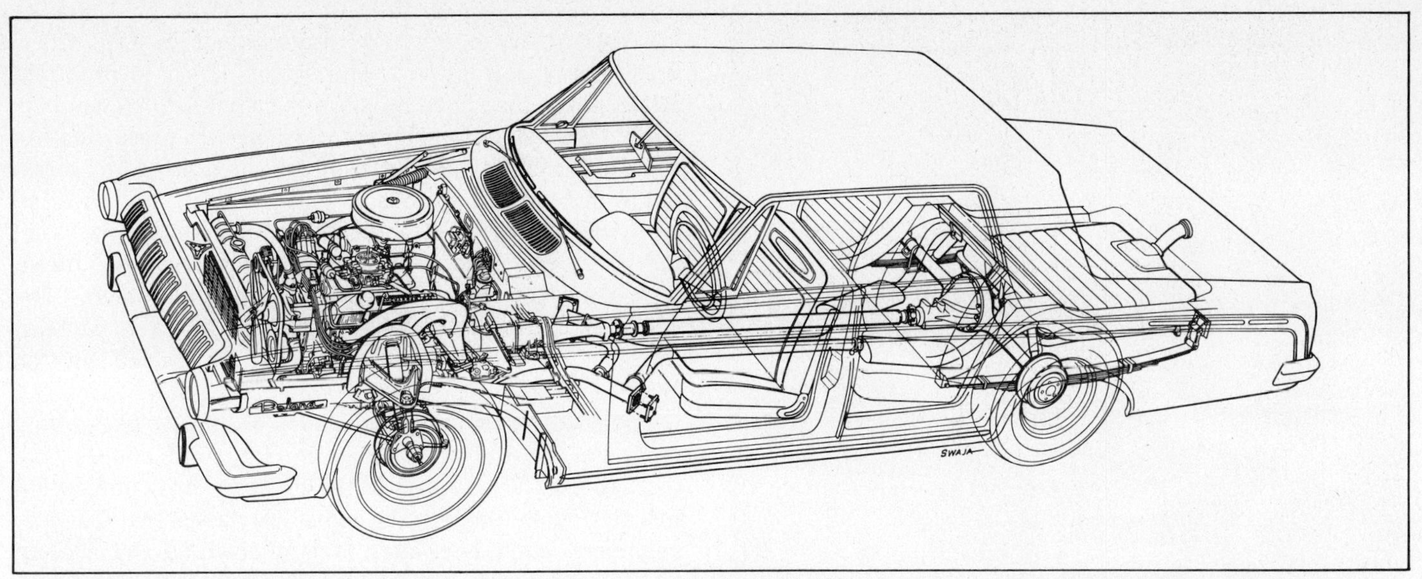

1963 Dodge 426 *Ramcharger* was a light Polara coupe with a complete big-block 426 race engine, cutout exhaust system, and heavy-duty suspension and drive line. It would run low 13s at 110 mph off the showroom floor.

looked like Swiss cheese. Another big chunk was eliminated by assembling the bodies without a drop of padding, sound deadener or insulation. The cars pounded like a bass drum going down the street. But who cared at the strip? The package was topped by putting a set of plexiglass windows in the back seat for the dealer to install!

Believe it or not Pontiac engineers chopped off another 300 lb! This brought the racing weight down to a little over 3300 lb. It was as good as adding another 50 HP. So good, in fact, that NHRA officials stepped in soon after the '63 Pontiac super-stock package was announced. They declared that all future factory race cars would be limited to a maximum displacement of 7 liters—about 427 cubic inches—and a minimum weight of 7.5-lb for each cubic inch of displacement. In effect, this created a 3200-lb minimum weight for a 427-CID engine. NHRA didn't want a lot of "Swiss-cheese" cars running their drag strips at 120 mph.

Dodge/Plymouth—A few months after the '62 Pontiac 421 was released, Chrysler fielded its own race-car lineup. The lineup included the Dodge Ramcharger and the Plymouth Super Stock models. The cars had very complete engine, drive-line and suspension packages. But no big effort was made to save weight, as the freshly downsized unit-body cars started out weighing less than 3400 lb with stock steel body panels.

The new race engine was based on the 413-CID B-block. Special parts included new cylinder-head castings with larger ports and valves—no exhaust-heat passages—forged 13.5:1 pistons, shot-peened rods, hard copper/lead main and rod bearings, high-volume oil pump, baffled oil pan, and a completely new valve train with a solid-lifter, 300° camshaft and dual valve springs. The only trouble with the dual valve springs was that valve-stem seals wouldn't fit. This resulted in '62 413s suffering from excessive oil consumption—as much as a quart every 50—100 miles. Still, the owner's manual said the engine was safe to 6500 rpm—but for sustained bursts not longer than 15 seconds.

Certainly the most-impressive things about the new

Dodge/Plymouth 413-CID Super Stock engine of 1962 had all the goodies, plus dual 4-barrel carbs on a cross-ram manifold. Good for 450 HP with open exhaust. Photo courtesy of Chrysler Corp.

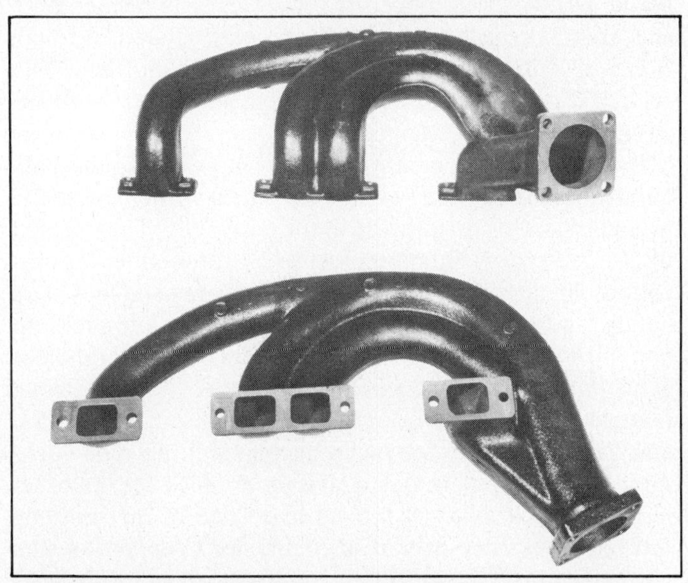

Special exhaust manifolds for the '62—'63 Dodge/Plymouth Super Stock engines looped up over the fenderwells, with passages split nearly full length. Those are 3-in. outlet holes!

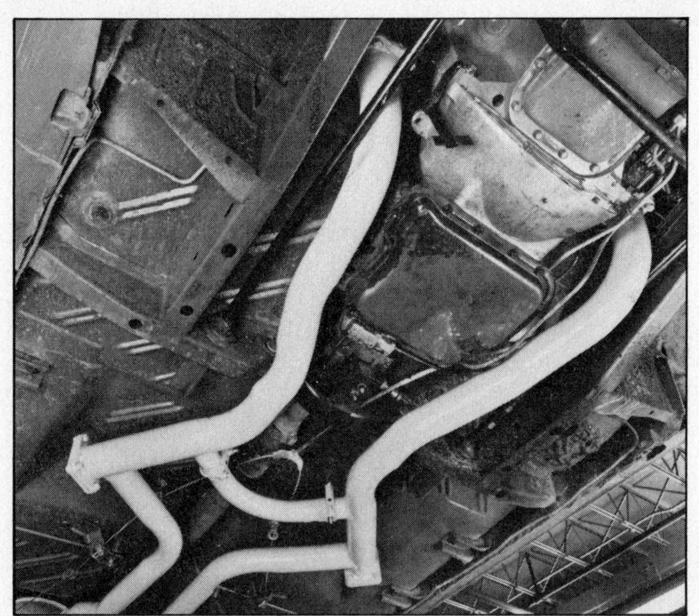
Undercar exhaust system on the Dodge/Plymouth Super Stocks was efficient for both street and race track. Big, 3-in. headpipes had cutout caps, with bypass pipes to dual low-restriction mufflers.

Chrysler race engine were the intake and exhaust manifolds. The intake was the unique short *cross-ram* I discussed earlier. Chrysler's cross-ram featured 15-in.-long manifolds criss-crossing over the engine and big Carter AFB 4-barrels mounted on integral plenum chambers above each valve cover. The layout was a natural evolution of the long-ram manifolds Chrysler had used in the '60—'61 period.

The new cross-ram aluminum manifolds were good for an extra 10—15 HP. And those exhaust manifolds: Huge iron castings looped up over the fenderwells to carry individual passages back to big 3-in. headpipes. There truly was nothing else like them in the industry.

This exotic exhaust system continued to the rear of the car. The headpipes were designed with dump tubes that could be quickly uncapped for street or strip racing. With the dumps capped, exhaust gas was routed through 2-1/2-in. tailpipes and dual low-restriction mufflers. The system was so efficient that the engine lost only 30—40 HP with the dump tubes capped!

It should also be mentioned that Chrysler was the only company that bothered to develop a heavy-duty automatic transmission for drag racing in the early '60s. They beefed up a three-speed, aluminum-case 727 Torqueflite. Modifications included: high-friction-coefficient clutches and bands, and recalibrated hydraulics for more line pressure and higher shift points. The result was a stab-and-steer drive line that was not only quicker out of the hole than a manual four-speed, but virtually bulletproof. Pushbutton gear selection made for quick manual shifting. The cushioning effect of the torque converter doubled the life of U-joints and rear axles. It proved to be one of the smartest racing moves they ever made. Chrysler Corporation cars cleaned up in all the top Stock/Automatic classes for years.

In 1963, Chrysler unveiled its *Stage II* racing package. It was a step forward in almost every area of the car. Bore was increased 1/16 in., upping displacement to 426 cubes. Breathing was improved with larger carbs, recontoured intake ports, machined reliefs in the combustion chambers around the valves, and a new camshaft with more duration and lift. Reliability at 6500 rpm was aided by adding material to the connecting rods around the bolts. Main-bearing grooving was increased threefold to increase oil flow to the rod bearings. And new valve-stem seals were fitted, remedying the excessive oil-consumption problem. The 426-II engines were 30—40-HP stronger than the '62 413s. Improved durability was especially appreciated on the NASCAR tracks.

The big story on the Stage-II body was the weight-cutting program. The unit-body Dodge and Plymouth coupes were the lightest full-size cars in the industry. But competition was getting so tough in the top-stock classes that Chrysler engineers wanted to bring weight down to the allowed 3200-lb minimum. All it took was an aluminum hood, inner and outer fender panels, radiator and bumper brackets. A secondary advantage of the new hood was that it gave an opportunity to design an air scoop to feed cooler, denser, outside air to the carbs. This scoop was said to add 2—3 mph to the trap speed.

These Stage-II Dodges and Plymouths are still winning on drag strips today.

Chevrolet—Chevrolet built a handful of *Factory Experimental* (FX) cars in mid-1962 with some aluminum body panels and basically standard 409-CID engines. But their first really serious factory-built drag-racing car came in early '63—the fabled *Z11* package.

The 409 block was stroked to give 427 CID—to take full advantage of the NHRA 7-liter limit. Breathing was improved radically with larger valves, recontoured ports, and a new type of *high-rise* intake manifold—raised 2 in. to reduce the curvature of the passages going into the cylinder-head intake ports. The higher passages also made it possible to leave an open area between the manifold and block to allow air to circulate and cool the intake passages. They called it an *air-gap* manifold. Ford and Chrysler were also experimenting with raised passages at that time.

Other refinements on the Z11 engine included stronger connecting rods with larger rod bolts, enlarged oil passages and a high-output oil pump, a higher-lift cam, and NASCAR exhaust manifolds with long, split primary passages. Detroit engineers were learning that exhaust passages should be separated for a few inches just like intake passages. This prevented overlapping pressure pulses in adjacent cylinders from interfering with smooth exhaust-gas flow. Aftermarket header manufacturers called this a *Tri-Y* arrangement. But to duplicate it in cast iron could result in a manifold weighing nearly 50 lb!

The Z11 package was based on the '63 Impala Coupe body. The Chevy racing budget couldn't support a radical weight-cutting program. But they did use an aluminum hood, inner- and outer-fender panels, bumpers and brackets to save 140 lb. Racing weight was just under 3500 lb.

But NHRA officials wouldn't allow the car to run in the Super-Stock class because only 20 or 25 cars were pro-

Ford and Mercury used the 427 *high-riser* engine for their factory super stocks and FX cars in '64—'65. Carbs were raised 3 in., which required a high bubble in the fiberglass hood.

duced. By this time, they considered 50 units as a reasonable minimum for stock classes. Before Chevrolet could do anything about it, the infamous no-racing edict came down from the GM front office in March, 1963: All GM divisions were officially out of racing for the forseeable future.

Ford 427 high-riser engines used a unique ram-air system. Large flexible tubes routed outside air from inner-headlight openings to an airbox above the dual Holley 4-barrels. Very efficient.

That handful of Z11 cars ran very successfully in the NHRA A/FX class, making a permanent name for themselves in drag-racing history.

Ford—Ford's drag-racing program was a kind of hit-and-miss affair in the early '60s. They were more interested in NASCAR racing anyway. In 1962, Ford did a radical weight-cutting project on a handful of 406-powered full-size Galaxies. The engineers whacked off some 500 lb by using aluminum, fiberglass and thin-gage steel throughout the bodies. Final racing weight was 3220 lb—one of the factors that brought on the NHRA 7.5-lb-per-cubic-inch rule.

The modifications proved to be extremely expensive. So, the following year, the weight-reduction project was set at a modest 160 lb. The '63 Galaxie-FX cars had a minimum of aluminum and fiberglass.

Ford's most-successful factory-built drag cars were the 1964 Fairlane Thunderbolts and Comet FXs. By basing the package on their intermediate-size cars, rather than the full-size Galaxies, they were able to get vehicle weight down to the minimum 3200 lb with little use of aluminum and fiberglass. Also, the shorter wheelbase helped weight transfer and traction. It was a real breakthrough for Ford—though Dodge and Plymouth had been racing for several years with similar-size cars.

The new 427-CID *high-riser* engine was also a vital part of the package. This was a *real* high-riser engine—not just a raised intake manifold, like the Chevy 409. Ford engineers also raised the intake ports 1/2 in. at the manifold face to get more flow area within the width restrictions caused by the pushrods. This increased manifold height that much more. Ford ended up with the carburetors *3-in. higher* than on a standard 427.

When provision was made for ducting to feed outside air to the carbs, there was no choice but to put a big raised bubble on the fiberglass hood. This wild teardrop-shaped hood bubble was certainly the most-visible feature of these cars. The driver could barely see over it. Needless to say, 427 high-riser Fords with bubble hoods were never put on the street. You may have seen some aftermarket imitations running around, but nothing Ford built.

Actually, 427 high-risers developed over 500 HP with tubular-steel exhaust headers. They had super-light, hollow-stem valves to allow 7000 rpm, with extra-high-lift camshafts. The cylinder-head ports and manifold passages were painstakingly developed on the first power-driven air-flow test bench in Detroit. For durability's sake, the cylinder block featured cross-bolted main caps and a main oil gallery along the *side* of the block that went directly to the main bearings—thus the name *side-oiler* 427.

And to top it, Ford made more than 50 Thunderbolts and Comet FX cars—61 to be exact—so the cars could run in the Super-Stock class. These cars easily dominated the manual-transmission S/S class in the '64 season.

Low-riser Ford 427 intake manifold (left) and a high-riser (right). As you can see, the name is derived from the slope of the *runners*—intake-manifold passages—and has nothing to do with cylinder-head port height. The high-riser required the famous teardrop hood bubble. The medium-riser was a more-responsive setup for the street. Read all about it in HPBooks' *How to Rebuild Your Big-Block Ford.* Photo by Steve Christ.

ABOUT HIGH-RISER INTAKE MANIFOLDS

There seems to be some confusion about the *high-riser* Ford 427 engine of 1963 and the *high-riser* intake manifold that appeared about the same time. Here's the scoop.

The Ford engine had the tops of the intake ports in the heads raised about 1/2 in. to give more port area without moving the pushrods to allow wider ports. These were termed *high-rise* ports. They required slightly higher runners, or risers, in the intake manifold. But Ford engineers went a step further. They raised these runners another 1-1/2-in. or so to reduce the curvature where the runners joined the head ports. This gave the intake charge a more-direct path into the heads and added horsepower at high revs. So the Ford was a *high-rise engine* that also used a *high-riser manifold.*

Where the confusion exists is that this new manifold design was in no way related to port design. An engine with low-height ports could benefit just as much from using a higher manifold that had less curvature going into the heads. In fact, Chevrolet hit on the idea for the raised manifold about the same time as Ford engineers. Chevy used one on the Z11 competition 427 in 1963. And the aftermarket manifold suppliers like Edelbrock began marketing the feature right away. Edelbrock manufactured high-riser manifolds for all popular V8s in the late '60s.

MEASURING 'EM UP

I was able to run drag-strip tests—or observe tests—on several of these factory-built race cars in the early '60s. In all cases, the cars were factory-engineering cars, or were fresh off the assembly line. No special tuning or modifications were made. The cars were just as you might buy one, except for 7-in.-wide slicks—maximum width for NHRA Stock classes at that time—and exhausts were unhooked or uncapped at the manifolds.

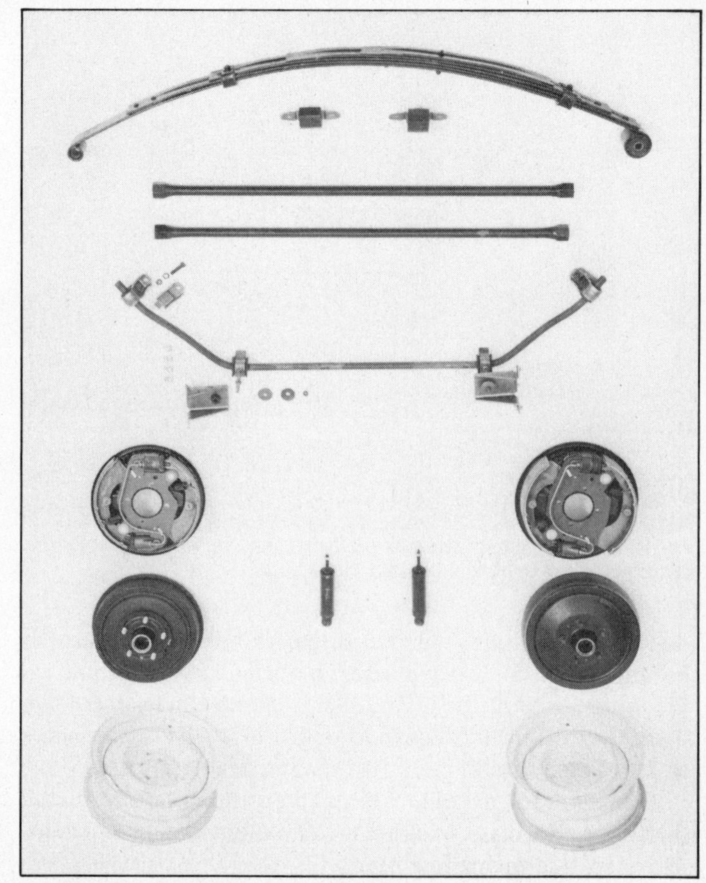

The late '50s saw the first widespread use of heavy-duty suspension components in high-performance cars. These are police parts used in Plymouth Furys and Dodge D-500s. Photo courtesy of Chrysler Corp.

Shown below, these figures wouldn't necessarily indicate which car would win in an organized drag event—after the usual professional tuning and legal modifications: tubular-steel exhaust headers, traction bars, and so on. All of the cars went considerably quicker and faster, even 20 years ago. And they'd be a lot quicker with today's suspension tricks and tires. But the figures do indicate that these factory-built race engines, with all factory equipment, could develop well over 400 HP with open exhaust. That's an honest 1 HP per cubic inch.

MORE THAN HORSEPOWER

There were other important supercar developments in

Car	Transmission	Axle ratio	1/4-Mile e.t. @ MPH
'62 Pontiac 421	M4	4.10:1	13.87 @ 108.43
'62 Dodge Ramcharger 413	A3	3.91:1	13.44 @ 109.76
'63 Chevrolet Z11 427	M4	4.56:1	13.18 @ 111.11
'63 Plymouth Super Stock 426	A3	4.30:1	13.11 @ 112.50
'64 Ford Thunderbolt 427	M4	4.56:1	12.91 @ 113.92

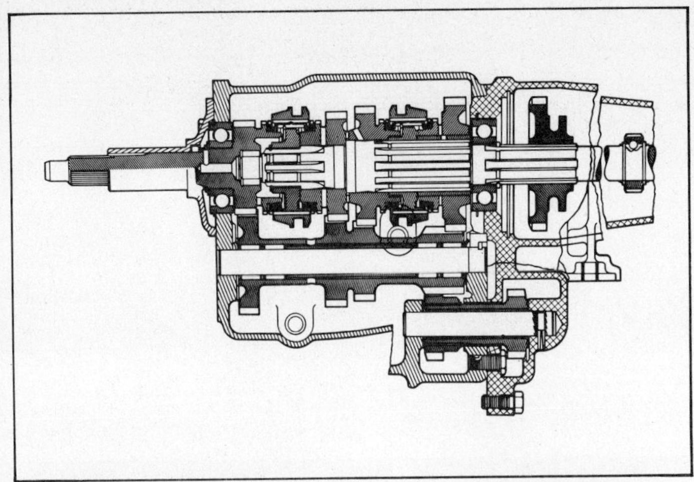

Borg-Warner T-10 four-speed was originally developed for the '57 Corvette. It was cobbled from the T-85 three-speed by putting reverse gear in the tailshaft housing. The tiny T-10 worked great in 283-CID 'Vettes, but when Detroit began putting it behind big-block engines, premature failures occurred.

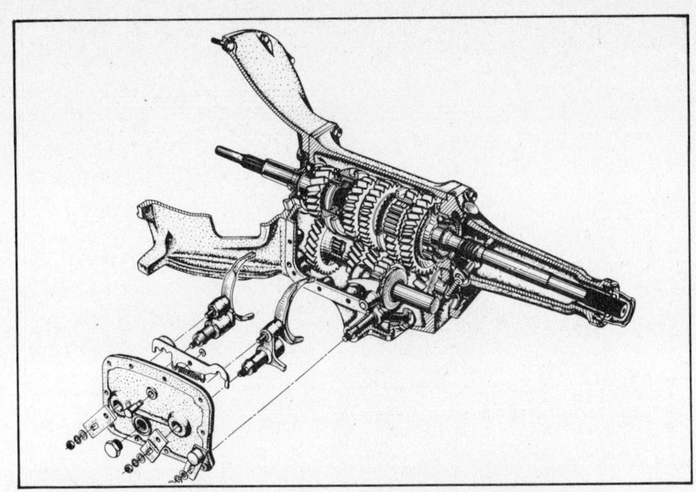

Chrysler's new A-833 manual four-speed transmission was introduced in 1964. It was much stronger than the Warner T-10 that had been used previously. It weighed 105 lb with cast-iron case. Drawing courtesy of Chrysler Corp.

the early '60s that established long-term patterns. One was the general use of heavy-duty suspension and brakes as mandatory equipment with high-horsepower engines. Actually, Chrysler first did this with their 300s, Plymouth Furys, Dodge D-500s and DeSoto Adventurers in the late '50s. They considered the heavy-duty chassis goodies as *safety* equipment. Any car capable of better-than-average speed and acceleration should be able to maneuver and stop better than average—like a police pursuit car.

Apparently, the other companies agreed. By the early '60s, most super-stock chassis were routinely fitted with high-rate springs and stiffer shock valving, harder, fade-resistant brake linings and sometimes a larger-diameter front anti-roll bar. Handling and braking were helped considerably—especially considering the nose-heavy weight bias with big-block engines. Street tires in those days didn't have enough bite to load the suspension and clutch. Even so, stiffer rear springs and shocks helped to control rear-axle windup and wheel-hop on hard takeoffs.

Another vital development area for supercars in the early '60s was the four-speed, floor-shift transmission. America's first heavy-duty, automotive four-speed was the Borg-Warner T-10. Developed by Borg-Warner and Corvette engineers in 1956, it was a quickie cobble job designed primarily for SCCA road racing against European sports cars with four-speeds. The new box used the gearcase and tailshaft housing of the Borg-Warner T-85 three-speed. Four forward gears were squeezed into the case and the reverse gears pushed back into the tailshaft housing. All four forward gears were synchronized. Shifting was done with a neat side linkage and short lever projecting through the Corvette console. It was tooled with minimum funds in only a few weeks.

Borg-Warner's new T-10 four-speed worked like a charm for the lightweight Corvette. It immediately became one of their most-popular options, especially for street driving. But when the transmission was released for two-ton Pontiacs and Chevrolets with big-block engines in 1959, problems began to crop up.

The problems weren't that hard to figure out. With four forward gears squeezed into a case designed for three, the gear teeth had to be narrowed. This weakened them. Heavy cars, high-plate-load clutches and high-torque engines played havoc. Most of the serious track and street racers used Warner T-10s in those days. They eventually became optional equipment on dozens of high-performance models from GM, Ford, Chrysler and even American Motors. But you had to be prepared to tear them down and rebuild at the drop of a gear. No one was satisfied with the T-10's durability in full-size cars.

It was inevitable that the Big Three would invest millions tooling up their own heavy-duty four-speeds in the '63—'64 period. These four-speeds were all much stronger than the T-10, though they shifted as quickly and smoothly and were not appreciably larger or heavier. The Chrysler A-833 was undoubtedly the beefiest of the bunch, followed closely by the Ford *top-loader,* with the GM Muncie M-21 the smallest and least strong. A variety of wide- and close-ratio gearsets were offered with these boxes, plus optional aluminum cases for some, and even different shifter styles.

These Detroit-built four-speeds added a new dimension to fun driving with high-performance American passenger cars. They remained a popular factory option for a decade.

KING OF THE STREET

If I had to name one showroom-available model that had a slight edge in performance in the '63—'64 period, it would probably have to be the 409 Chevy—with the 427 Ford only a whisker behind. Ford had more exotic goodies for the 427 engine, but most of them were not used in the street 427s. Chevy offered more-aggressive pieces in the street 409s—though the anti-racing edict of March, 1963 slowed them down. Chrysler and Pontiac directed most of their performance budgets toward racing cars. As a result, the 1963 anti-racing edict hit Pontiac harder than Chevrolet.

It hit them so hard, in fact, that they turned to an entirely new concept in high-performance, youth-oriented cars.

The car that really got the supercar movement going—the GTO. This is a '66 model. Photo by Ron Sessions.

Pontiac Invents the Supercar

The GTO legacy

A NEW LOOK IN STREET MACHINES

The cars I talked about in Chapter 3—the big-cube factory super stocks of the early '60s—were great straight-line performers. And they were a lot of fun to drive, with four-speed, floor-shift transmissions and heavy-duty suspensions. I will even say that these cars were *very* responsive. But they left a whole lot to be desired as practical, everyday transportation.

Their engines were just too highly tuned. Radical cam timing, huge ports and valves, big manifolds and copious carburetion were a terrible combination for normal cruising on the street or highway. I'd describe the engines as noisy, rough and hard to start. They had a lumpy idle—like a race car—and poor throttle response at low speeds. Worst of all, gas mileage dropped as low as 6–8 mpg in normal stop-and-go driving. Cold-weather driving was nearly impossible, because some engines didn't even have exhaust-heat risers in the intake manifold.

Performance enthusiasts of the day were well aware of these faults. Only a few hardy souls ever actually tried to drive 409 Chevys or 406 Fords on the street every day. Annual sales of these cars never totalled more than 1500 or 2000 units—only a drop in the bucket, even 20 years ago. Those early super stocks were highly respected and held in great awe by the street crowd. But they never made a dent in the potential *youth/performance* market. And never really made any money for the companies that produced them on a per-unit basis. In fact, they probably *cost* the companies money. The early-'60s super stocks were strictly image-makers.

Above: Early GTOs had clean, understated lines. Photo by Ron Sessions.

It took Pontiac to find a new formula for making money on street performance. The 1964 GTO was the car that turned everything around. I'm calling this the first real *supercar*—not because it was any more super in performance or handling than some of the super stocks that had gone before. Not because it had anything new in the way of equipment or design. It was the *concept*—the idea that made it different. It was the way existing equipment was combined into a new *package*—with a new image—that made the GTO special.

It's hard to pinpoint credit for the GTO idea. But I believe the two men most responsible were John DeLorean and Jim Wangers. DeLorean was chief engineer for Pontiac at that time. He was a man of tremendous drive, with a magnetic, persuasive personality. In the '60s, he had a lot of influence on Pontiac's general manager, Pete Estes. DeLorean was well aware of the tremendous impact performance and image had on Pontiac sales in the late '50s, and he was ready to run in any direction to keep the momentum rolling through the '60s.

Jim Wangers was equally enthusiastic about performance, Pontiacs and profits. He was an account executive at MacManus, John & Adams—Pontiac's advertising agency. His job was dreaming up catchy advertising themes for the Poncho performance models. He also acted as a liaison be-

Pontiac's GTO was a big sales success. Package included big-displacement V8 in a lightweight intermediate-size coupe, with heavy-duty suspension and some special trim. Photo courtesy of Pontiac.

tween Pontiac promotion, product and advertising people. Wangers was a close personal friend of DeLorean. The two thought on the same wavelength where performance was concerned. Between them, they conjured up all kinds of wild ideas to promote Pontiac to the youth market. DeLorean had the engineering know-how to turn ideas into hardware. Wangers had a fantastic flair for promotion—not only hardware, but an idea—a dream. His ads made a GTO appear to be the most-desirable car in the world.

I might also mention that GTO promotion wasn't a nine-to-five job with Wangers. He lived and breathed GTOs and Firebirds for five or six years. I can recall many a day spent in a garage or on a windswept drag strip with Wangers and a crew of mechanics, testing different equipment combinations or gear ratios on factory-prototype cars. He always wanted a writer and photographer on hand so any *good* results would get into print! And he always had a perfect excuse when something went wrong! He was as sharp at generating *ink*—magazine publicity—as he was in putting customers in the mood to buy Pontiacs.

The DeLorean/Wangers team practically assured something revolutionary when the first GTO was announced in the fall of 1963.

THE GTO CONCEPT

There wasn't anything very complicated about the DeLorean/Wangers GTO concept. They took an age-old performance principle—high power-to-weight ratio—dressed it up to appeal to young street cruisers, and made sure the whole package was smooth, quiet and economical enough for everyday driving.

Look at it this way: The super stocks of the early '60s had been based on full-size cars. After all, most people bought full-size cars. And super-stock-size engines would not fit in compacts anyway. But it meant lugging around as much as 4000 lb.

In 1961, the GM divisions introduced a new line of *intermediate*-size cars: Olds F-85, Pontiac Tempest and Buick Special. A few years later, Chevrolet followed with the Chevelle. They had a wheelbase of 112 in. and a curb weight closer to 3000 lb. So, Delorean and Wangers thought, instead of putting a highly tuned big-block V8 in a 4000-lb car, why not put a *mildly* tuned big-block in a 3000-lb car? Engine output might drop 70—100 HP with a milder cam, smaller carb, and so on. But the package was also shedding up to half a ton of dead weight. In the final equation, the horsepower-to-weight ratio ended up *about the same* as the big-cube super stocks. This meant equal acceleration with a smoother, quieter, more-flexible engine capable of twice the fuel economy of the super stocks.

So much for theory. A lot of thought was given to projecting a new *image* with this new type of performance car. Realize that the super stocks of the early '60s were practically devoid of any special external identification. Maybe a tiny fender medallion with a 409 or 406 under crossed flags. But nothing in the way of special body trim, wheel designs, instrument clusters, grille emblems, hood designs, or the like. You had to squint twice to see if you were getting blown into the weeds by a 409 Chevy.

When you bought the 409 package, you were paying for an engine and heavy-duty suspension—not a lot of special body goodies. Wangers didn't want this for the GTO. He

To reinforce the European character of the GTO name, Pontiac indicated engine displacement in metric units—6.5 liters, or litres, equals 389 cubic inches. Photo by Ron Sessions.

This was one way to tell a GTO from a bread-and-butter Tempest. Photo by Ron Sessions.

1964 GTO sported a blacked-out grille. Photo by Tom Monroe.

wanted *everyone* to know they were seeing a GTO. But he didn't want to hit them over the head with it. Identification must be positive, but subdued. No 10-in.-wide decals splashed all over the side—but more than just a 3-in.-long fender emblem.

What evolved was a masterpiece of subtle image-making. The front view of the car—which any stylist knows is the most-important first impression—was set off by merely blacking out the standard grille bars and putting a simple GTO emblem in the opening. Similar GTO emblems went on the back fenders and deck lid, and there was a special checkered-flag medallion behind the front wheels. The cosmetic package rounded out with two subtle, non-functional air scoops on the hood, sporty wheel covers and red-stripe premium tires to accent the lower half of the car. That was it. Nothing fancy, but *positively GTO.*

Suspension calibration was unique to GTO, too. But it was done in a more-sophisticated way than the super stocks of the early '60s. Spring rates, shock valving and stabilizer-bar stiffness were carefully matched to the front/rear weight balance and cornering capabilities of the wheel/tire combination. The object: to give really outstanding cornering power and high-speed stability.

Some magazine testers compared the GTOs handling with expensive European sports cars. It was voted the American car-of-the-year in several polls, where special emphasis was on handling. Often mentioned was the excellent high-speed braking with the optional sintered-iron linings. These brake linings had to be warmed up to work well, but after that there was no such thing as fade. One magazine clocked a Tri-Power GTO from a standing start to 100 mph and back to a dead stop in an incredible 19.4 seconds! That was equivalent to the performance of a $25,000 Ferrari of that period.

How about horsepower? The engine was surprisingly tractable, considering its performance. Essentially, Pontiac took a standard 389-CID high-compression 4-barrel engine and installed a slightly higher-output hydraulic-lifter camshaft and low-restriction air cleaner. Exhaust was routed through a dual system with special low-restriction mufflers that gave the right burbling sound! That's all! It produced about 20-HP more than the standard 4-barrel

On Ram Air IV-equipped '70 GTOs, air was ducted through wide scoops at the leading edge of the hood. Photo by Tom Monroe.

engine. But it certainly wasn't any wild semi-race engine. It didn't need to be.

One more vital ingredient in the new GTO formula—at least as far as DeLorean and Wangers were concerned—was an arm-long list of optional equipment. This was a list of more than just fancy upholstery and courtesy lights. There was all-out performance and image stuff. Here's a sampler: four-speed floor-shift transmissions with close-ratio gears, three-2-barrel carburetion, finned aluminum wheels, metallic brake linings, tachometer, limited-slip differential, a broad choice of final-drive ratios, extra-stiff suspension and oversize tires. These extra goodies not only allowed the buyer to customize his GTO on the order blank; it meant huge profits for the factory and dealer. Most GTOs were sold with at least $1000 worth of extra equipment.

It's history now that Pontiac's revolutionary '64 GTO was one of the great auto-success stories. The way sales skyrocketed astonished even the most-optimistic planners. In fact, Pontiac planned to make only about 5000 cars in '64—enough to test the market. But by January, barely into the model run, dealers sold 10,000 of them, and were happily compiling waiting lists. By the end of the '64-model run, 32,000 "Goats" had gone out the door—

Ram-Air was available for the first time on the '66 GTO. Photo courtesy of Pontiac.

GTO stands for Gran Turismo Omologato—not Gas, Tires and Oil. Photo by Ron Sessions.

and Pontiac dealers were screaming for a new, flashier and more-expensive GTO for 1965.

The GTO set the pattern for the American supercar era. Within two years, practically every company had its own GTO clone on the streets. Here's a brief rundown of how the various companies approached this dynamic new market segment....

Oldsmobile—I give credit to Oldsmobile sales people for being the first to act on the GTO challenge. Within weeks of the GTO's showroom debut, Olds engineers were finalizing an option package for the F-85 Cutlass intermediate coupes that would try to match the "goat" in both acceleration and handling.

Actually Olds already had the raw material for an intermediate supercar in their '64 parts catalog—the *police pursuit* version of the Cutlass coupe. This had heavy-duty suspension and brakes, and used a special high-output version of the new 330-CID low-block. Goodies included a 4-barrel carb, high-compression pistons, long-duration cam, dual exhaust and a dual-snorkel air cleaner. All Olds had to do was throw in a Muncie M-21 four-speed transmission, a few identification emblems—and they had their famous 4-4-2 performance model. Those numbers stood for 4-barrel carburetion, 4-speed transmission, and 2 exhaust pipes.

Olds had the car in showrooms by April of 1964, only seven months after the GTO appeared. Though many purists will say the '64 4-4-2 was a *parts-bin special,* it proved to be a decent performer.

Those early 330-CID 4-4-2s performed better than the pounds-per-cubic-inch factor would suggest. But engineers and sales people alike all agreed that Olds should take full advantage of the 400-CID allowance for GM intermediate models.

In 1965, the 4-4-2 got a new V8. The block deck height was raised and the engine was bored and stroked to 400 CID. To handle the longer rods and increased stroke, bigger bearings were used. The 400-CID 4-4-2 pulled

Oldsmobile had their 4-4-2 package in the showrooms only six months after the GTO appeared. It was essentially a 330-CID police-pursuit engine and police suspension in a plain F-85 coupe body. Photo courtesy of Oldsmobile.

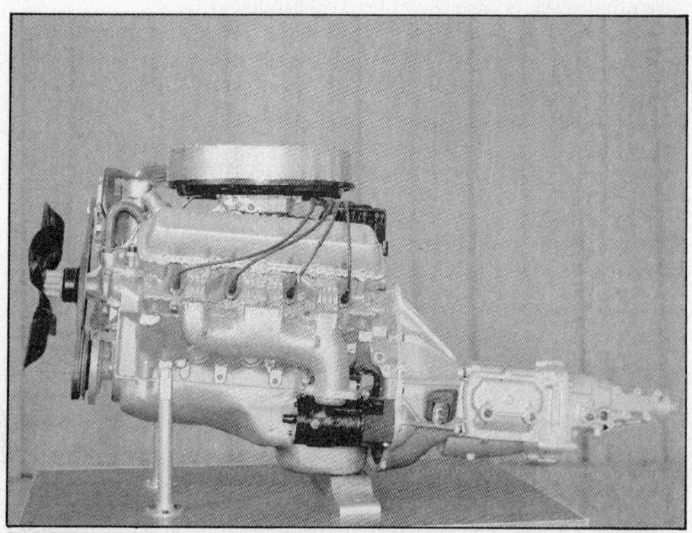

Big Olds 400-CID, 4-barrel V8 hooked to a GM Muncie four-speed manual transmission was the popular powertrain for Olds performance in the late '60s. Up to 300 net HP with optional W30 cam. Photo courtesy of Oldsmobile.

A bit of trivia here on the '64 330 CID cars, 4-4-2 stood for *4*-barrel carburetor, *4*-speed transmission and *2* exhausts; when displacement was bumped to 400 cubic inches for '65, Olds' ad agency claimed that one of the "4"s stood for *400* CID. Photo courtesy of Oldsmobile.

Though '64 4-4-2s with the 330-CID engine were decent performers, the '65 400-CID engine really woke up performance. Photo courtesy of Oldsmobile.

10—20 HP more than the 330 at the top end. But most important, the 400 offered more mid-range torque and sharper throttle response.

The Olds 4-4-2s of the mid-'60s were noted even more for superior handling. They had a gimmick: an anti-roll bar on the *rear* suspension as well as on the front. This caused the rear tires to handle more of the total cornering force and allow a bigger front bar, giving the steering a more-*neutral* feel. The front end didn't *push* as much in the corners. You could hold the car in a four-wheel drift with light application of the throttle. Most other manufacturers avoided rear-anti-roll bars because they thought the cars would be too twitchy for the average American driver. Olds dared to be different—and their 4-4-2 was the best-handling of all the early supercars as a result. It's significant that some of the other companies supplied rear-anti-roll-bar kits for their muscle cars!

It's also worth noting that Olds sold more than 30,000 4-4-2s in 1968. It was a popular car in the supercar era.

Buick—I haven't talked much about Buicks since the dual-carb jobs of 1941. There's good reason. Throughout most of the late '40s, '50s and early '60s, Buicks were known as old-men's cars. When sales were good, Buick was content with the image. But when the numbers started falling, Buick looked to the rise of the Pontiac Division and its new performance image. It was time to act.

Buick's first *Gran Sport,* or GS model, of 1966 used a standard 4-barrel version of their old 401-CID *nail-valve* V8. Photo courtesy of Buick.

Buick performance improved substantially with a totally new 400-CID engine in 1967. This is the '68 Buick GS 400. Photo courtesy of Buick.

Chevrolet offered several power-pack versions of the Mark IV big-block in the late '60s, in 396, 402 and 427-cubic-inch displacements. Most used standard heads, manifolds and carburetion, with high-lift hydraulic cams. Photo courtesy of Chevrolet.

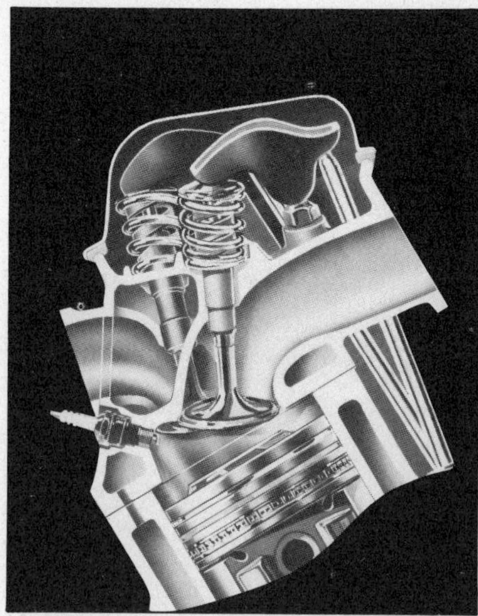

Unique *porcupine* valve layout of the Mark IV Chevy gave unusually good breathing by reducing the curvature of ports and the shrouding around the valves. Port and valve sizes were generous. Photo courtesy of Chevrolet.

Still, it took Buick two years to see the light and get their GTO replica on the streets. In 1966, they introduced the *Gran Sport,* or GS, version of the Skylark intermediate coupe. The car came with special trim and wheels, a 401-CID engine, heavy-duty suspension and all the right goodies to appeal to the new market. About the only thing they omitted was a special, high-output camshaft for the engine. Buick just stuck in their standard high-compression 4-barrel engine. This did compromise performance a bit.

This original 401-CID GS engine was basically the old *nail-valve* V8 with pentroof combustion chambers, which had been in production more than 10 years. It had good intake porting, but the exhaust side had tiny valves and sharply curved ports. It was early-'50s engineering. Performance was not outstanding. But in 1967, Buick introduced an all-new 400-CID big-block V8 with in-line valves, rounded chambers, large ports and stable valve gear. Performance improved considerably.

Chevrolet—Chevy sales people started scrambling when it was obvious Pontiac's new GTO was a money-maker. Chevy had all sorts of high-output small-block V8s released for Chevelle Super Sport models from the very start—but no big-blocks! The '64 Chevelle SS with its 10.5:1 compression, 300-HP 327 was no slouch. And the '65 Chevelle SS was available with a very potent 350-HP 327 with 11:1 compression. But the public wanted big blocks!

Midway through the '65 model year, Chevy introduced the new Mark IV 396-CID big-block engine. The best they could do for 1965 was to get 200 hand-built Chevelle SS 396s into the hands of a few VIPs and press people. The Super Sports had some special trim and other goodies. The 396-CID Mark IV *semi-hemi* had big-port heads, Holley carb, and a special long-duration, hydraulic-lifter cam. This particular engine combination never went into production, and the handful of '65 SS Chevelles with the Z16 option are priceless collectors' items today.

Chevrolet's supercar entry appeared in 1966—a Chevelle coupe with standard 396-CID Mark IV engine, heavy-duty suspension and special trim. They called it the *SS 396.* Photo courtesy of Chevrolet.

The real McCoy came in 1966. With fresh sheet metal gracing its flanks, the '66 Chevelle SS 396 was the GTO-class contender Chevy fans were waiting for. It had all the GTO ingredients, including a special high-lift, hydraulic-lifter cam that raised gross output from 325 HP to 350 HP at 5200 rpm. It also featured a dual-snorkel air cleaner and dual exhaust. Otherwise it was a standard 396 4-barrel engine. Transmission options included manual three-speed, four-speed or the venerable two-speed Chevy Powerglide. The bulletproof three-speed Turbo Hydramatic 400 was to come later. A choice of axle ratios, police suspension, various tire sizes and many other options permitted the buyer to dream up his own custom SS on the order blank, just like Pontiac dealers had encouraged GTO buyers to do.

Within two years, SS 396 sales climbed to more than 60,000 units a year.

Ford offered a special manual-shift automatic in their GTA supercars of 1966. It proved to be more popular than the four-speed GT versions. The GT version of the 390-CID V8 gave good acceleration. Photo by Milo Rowell.

Mercury's answer to the GTO was the Cyclone GT. This is the interior of a '67 Cyclone convertible. Photo by Ron Sessions.

GM Top-Brass Worries—The Chevy, Olds and Buick permutations on the new Pontiac GTO theme started some head-scratching at the GM front office. Top brass immediately began questioning how the new high-powered, lightweight cars were squaring with their anti-racing edict of 1963. There was fear that safety critics would zero in on these cars—especially if they started pouring out of all the divisions—and GM could quickly lose all the good PR their new stance had generated. On the other hand, the big boys didn't want to choke a goose that appeared to be laying golden eggs—lots of 'em.

They ended up compromising. Big-block V8s were OK in intermediate-size cars, but not over 400 cubic inches—401 in Buick's case. In the late-'60s, they eased this policy somewhat. The new formula was this: Not less than 10 lb of curb weight for each advertised gross HP. This permitted using bigger engines in the heavier intermediates, some of which gained 400—500 lb from the mid- to late '60s.

Ford/Mercury—As with Chevrolet, Ford offered their potent, high-performance small-block V8 in intermediate-size cars—the Fairlane and Comet—as early as '64. The 271-HP, 289-CID was a real stormer, but no big-block. The big-block FE engine wouldn't fit in these cars without extensive modifications.

Ford responded to the GTO challenge in 1966 with Fairlane and Comet *GT* and *GTA* models. These were based on their freshly upsized 116-in.-wheelbase, intermediate unit-body cars. They used the 390-CID version of the FE engine, boosted to 335 HP with a high-lift cam, 600-cfm Holley carburetor and dual exhaust system.

One unusual feature of the Ford/Mercury GTA models—the "A" stood for automatic—was an optional *Sport Shift* three-speed automatic transmission. It could be shifted manually with a neat console-mounted T-handle. Positive detents in the shift gate made manual upshifting more precise.

Ford unleashed a potent supercar combination by using the 428 Cobra Jet engine in the '69 Torino coupe to make the Torino Cobra. It was considerably stronger than the former 390-CID Fairlane combination. Photo courtesy of Ford Motor Co.

Also, 1966 marked the introduction of Ford's heavy-duty C-6 automatic transmission for use behind high-performance big-block engines. The C-6 proved so tough and the Sport Shift linkage so popular that most Ford dealers ordered the bulk of their GT inventory with it. Only a handful of *top-loader* four-speed GTs were ever produced. This compromised the performance image of the cars because the manuals were considerably quicker than the automatics.

Ford completely retooled their intermediate cars in 1968, going to a still larger, longer body. This added about 300 lb to the coupe. As could be predicted, performance of the GT models with the original 390-CID engine was mortally wounded. The following year, Ford released the new *Cobra Jet* version of the 428-CID FE block. Ford dropped it into the Torino fastback coupe, making a new *Cobra* model to do battle in the boiling GTO market. The former

By '69, Mercury was offering their largest big-block, the 428 CJ, in the Montego GT. This one was equipped with ram-air induction. Photo courtesy of Ford Motor Co.

"GT" designation was dropped and forgotten.

The new 428 CJ engine was half again as strong as the 390 GT. It used the big-port cylinder heads from the 427 NASCAR medium-riser engine, a 735-cfm Holley carb with optional *ram-air* induction, new hydraulic-lifter cam and high-rpm valve train, and low-restriction exhaust manifolds. The new 428 CJ carried the same power rating of 335 HP as the 390, but in a more-convincing way. The cars were very successful in drag racing. It was perhaps Ford's best compromise between brute horsepower and street driveability in the entire supercar era.

But strangely enough, those 428-CID Torino Cobras never sold half as well as the earlier 390 GTs. There were several reasons. Some disappointed GT buyers were afraid to take another chance. Many had switched over to GM and Chrysler muscle cars. And many had turned to big-block Mustangs. But even when Ford provided a stronger base engine for the 1970 Torino Cobra, sales failed to rally.

The stronger engine I'm referring to was the Cobra Jet version of the 429-CID *385-Series* V8. It featured staggered valves and *semi-hemi* combustion chambers—like the big-block Chevy. This engine also had special big-port cylinder heads—30% larger than the 428 CJ ports—big Holley carb, an even higher-output camshaft, individual ball-joint rocker arms, and an excellent dual exhaust system. This was one of the few Detroit supercar engines that had almost too much port and valve area for responsive street performance. But with the right spark timing and gear ratio, these '70 Torino Cobra 429s were hard to beat from 40 mph up.

Chrysler waited until 1967 to bring out their supercars, then fooled the industry by using 440 cubes instead of 400. This is the '67 Plymouth GTX.

Chrysler based their supercars on the intermediate B-body—which was slightly larger than some of the GM cars. This is the '67 Dodge Coronet R/T model. Photo courtesy of Mopar Muscle Club.

But alas, so few of them were sold that many supercar fans have never even seen one.

Dodge/Plymouth—The Chrysler divisions offered intermediates with high-performance big-block engines all along—not to mention the first street Hemi of 1966. But Dodge and Plymouth didn't get their GTO copies into production until three years after the market broke. The 1967 Dodge Coronet R/T and Plymouth Satellite GTX, though, proved to be worth waiting for. Remember that the Chrysler B-body intermediates were a bit larger and heavier than equivalent GM models. With this in mind, and not deterred by a front-office policy that limited them to 400 CID, Chrysler engineers based their new mass-produced supercars on the big 440-CID wedge. This engine had only recently been introduced for heavy luxury cars.

Chrysler didn't just stick in a high-performance cam and let it go at that. The 440 engine was completely re-engineered for use in high-performance street cars. This included new big-port cylinder heads, big-passage intake manifold, 650-cfm Carter AVS 4-barrel, high-output cam and high-rpm valve gear. The package also had low-

Big 440-CID Magnum engine used in Dodge and Plymouth muscle cars in the late '60s was another example of re-engineering for performance. Chrysler developed new cylinder heads for it. Photo courtesy of Chrysler Corp.

Pontiac abandoned the fabled Tri-Power carburetion, left, for a single 700-cfm Quadrajet in 1967. The Q-Jet gave just as much power, and was a lot simpler and cheaper. Ram-air bonnets are behind each system.

restriction exhaust manifolds, hard rod and main bearings, and a windage tray—plus the usual low-restriction air cleaner and dual exhaust system. And a declutching radiator fan saved another 10 HP above 3000 rpm.

These Dodge and Plymouth supercars, though a little heavier than some equivalent GM and Ford models, were perhaps the strongest of the bunch in the late '60s. They developed a reputation for having a bulletproof drive train. Street enthusiasts rewarded the effort by buying some 17,000 Plymouth Satellite GTXs in 1968. Dodge dealers did less well, never selling more than 7000 Coronet R/Ts per year.

In '68, the R/T name was affixed to the performance version of the Dodge Charger. The Charger had unusually clean *fuselage* styling. With the same mechanicals as its Coronet-based R/T sibling, Charger R/Ts sold very well for several years.

Pontiac—I've already discussed the original 1964 GTO in

Dodge's famous *Scat Pack* of 1968 consisted of the Dart GTS 340 (top), Super Bee 383 (middle), and the Charger R/T 440 model (bottom). Bold bumble-bee stripes around the hind quarters identified them. Photo by David Gooley.

Jim Wangers dreamed up the GTO *Judge* model in 1969, named after a routine on the popular TV show, Laugh-In. Early models were painted bright orange, with wild spoilers and decals. Valuable collector cars today. Photo courtesy of Pontiac.

The Judge decal set these cars off from regular GTOs. Photo by Ron Sessions.

1968 GTO scooped the industry with the first use of an integral energy-absorbing front bumper. *Endura* bumper was standard, but could be deleted in favor of the chrome LeMans bumper. As a result, GTOs with the chrome front bumper are very rare. Note the hood-mounted tachometer pod. Photo courtesy of Pontiac.

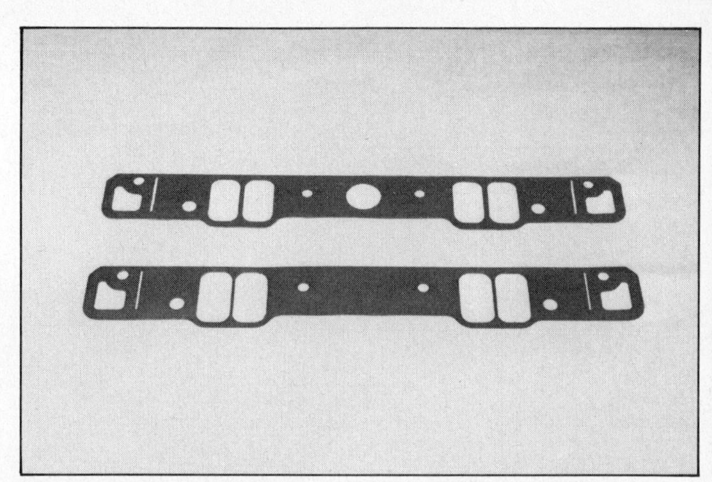
Pontiac supplied special intake-manifold gaskets in the '60s to block off all exhaust heat to the intake-manifold crossover passage. Blocked gasket shown at bottom. This added up to 10 HP on a warm day. Photo by Wally Chandler.

Jim Wangers was also behind the American Motors *Rebel Machine* in 1970. It was a dolled-up Rebel coupe with a tweaked 340-HP 390-CID V8. Photo courtesy of American Motors.

High-visibility graphics of the '70 Rebel Machine were a sign of the times. Photo courtesy of Hurst Performance, Inc.

some detail. The next major change in the combination came in 1967, when the basic Pontiac V8 was extensively retooled for better performance and durability. Changes included: new cylinder-head castings with larger ports and valves—2.11-in. intakes and 1.77-in. exhausts—new intake manifolds, a switch to the 750-cfm Rochester Quadrajet for all 4-barrel engines, new camshaft, and a reinforced block with thicker main webs. A bore increase upped displacement from 389 to 400 cubic inches. Street performance was helped considerably. And high-winding race cars benefited from the larger ports and valves, which added 10—15 HP in the 5000—6000-rpm range.

Incidentally, after '66, the Tri-Power option was quietly laid to rest because the new Quadrajet flowed almost as much air as the three 2-barrels without the linkage headaches.

Pontiac never relinquished the head start they got when they scooped the industry with the '64 GTO. With Jim Wangers pounding away in the media and DeLorean's engineers grinding out new performance and image gimmicks month after month, GTO sales soared ahead of the industry. They hit almost 100,000 units in 1966. Sales remained above 70,000 annually to the end of the decade. No other company was close. The GTO was one of Pontiac's top profit-makers in the late '60s. No other high-performance production car ever approached its popularity.

American Motors—Even tiny AMC tried to wrest itself from the stodgy, George Romney-era economy-car image and cash in on the '60s supercar craze. Considering the company's comparatively meager resources, its supercar offerings didn't fare badly.

American Motors concentrated its performance efforts on the Javelin and AMX ponycars in the late '60s. The only real GTO-type model they made was the curious *Rebel Machine* of 1970. Essentially, the package was a hopped-up 390-CID V8 in their 114-in.-wheelbase intermediate coupe—with plenty of special trim and styling gimmicks for identifiction. With its red, white and blue paint job, the car looked like an American flag going down the road.

Many car fans didn't realize that there were a lot of specially designed and tooled mechanicals under the skin.

Buick went to 455 cubes for their GS supercars in 1970, after the GM corporate ban on big displacements was dropped. The extra displacement just about offset the cars' increased weight. Photo courtesy of Buick.

These included a big-passage intake manifold with 600-cfm Autolite 4-barrel, high-lift cam, high-rate valve springs, better-flowing exhaust manifolds and low-restriction dual mufflers. Backing up the potent power plant was a Borg-Warner four-speed transmission, Hurst competition shifter, 3.54:1 final drive with limited-slip and heavy-duty suspension. External stuff included mag-style wheels, white-lettered tires, functional hood scoop and hood-mounted tachometer. It was really a lot of car for the price.

And performance was competitive. The owner's manual—*that's right, the owner's manual!*—claimed quarter-mile times of 14.4-second e.t. at 98-mph trap speed. Not bad for a 3700-lb car. The sad truth about the Rebel Machine was that American Motors products didn't have a strong image with the supercar subculture in the late '60s. Few would've bought them regardless of how fast they were. Only a handful survive today.

ACCELERATION PERFORMANCE

It's time to break out the published road- and drag-strip tests on these GTO-class cars, and see how their performance compared. Keep in mind that these are all strictly standard models—no optional engines. It should also be mentioned that I'm not including *all* of the CAR LIFE tests published during these years. In a few cases, the test car was obviously *doctored*—specially tuned—before the magazine testers got it. In these cases, acceleration figures were not representative of what you could buy in the showroom. I've omitted these tests so as not to confuse the issue.

Below is a sampling of what I consider the most-representative CAR LIFE tests.

You can see that the typical performance of these standard GTO-type cars of the late '60s was similar to the factory super stocks of the early '60s. And yet, these later cars were much smoother and more suited to street driving. The secret, as pointed out earlier, was using a mildly tuned engine with lots of cubic inches in a comparatively lightweight intermediate body. The power-to-weight ratio—and thus the acceleration—ended up about the same as the earlier super stocks. Mild camming and carburetion assured smoother, more-economical everyday operation.

But it's still obvious that cubic inches counted for a lot, even in the milder configurations of the GTO-type cars. Note that the 440-CID Plymouth GTX was the quickest of the listed cars—while the 428-CID Torino Cobra was next. By the late '60s, the 400-CID GM intermediate supercars were at a distinct disadvantage. That situation was remedied in 1970 with a new generation of 454- and 455-CID big-blocks.

POWER PACKS FOR POWER PACKS

Anyone could have predicted that the standard, mildly

Car	Transmission	Axle ratio	Adv.HP	0—30 mph	0—60 mph	1/4-mile e.t. @ mph	Est. Net HP
'64 Olds 4-4-2 (330-CID)	M4	3.23:1	310	3.0 sec	7.4 sec	15.6 @ 89	240
'65 Olds 4-4-2 (400-CID)	A3	3.23:1	345	3.3	7.8	15.5 @ 89	250
'66 Buick GS	A2	3.08:1	325	3.0	7.4	15.3 @ 88	240
'66 Ford Fairlane GTA	A3	3.25:1	335	3.6	8.6	15.4 @ 87	230
'66 Pontiac GTO	M4	3.08:1	335	2.8	6.8	15.4 @ 92	270
'68 Olds 4-4-2	A3	3.42:1	350	3.0	7.0	15.1 @ 92	280
'68 Plymouth GTX	A3	3.23:1	375	3.0	6.8	14.6 @ 96	330
'69 Ford Torino Cobra	M4	3.50:1	335	3.2	7.3	14.9 @ 95	310
'70 Chevelle SS 396	A3	3.31:1	350	3.3	8.1	15.5 @ 90	280

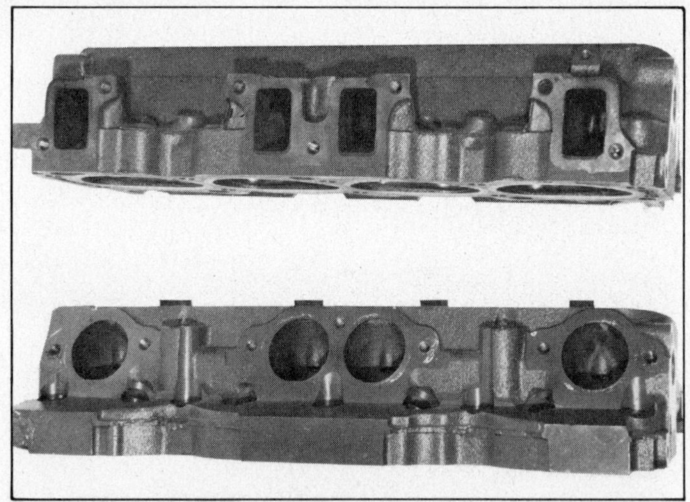

Buick developed special *Stage II* cylinder heads for drag racing in 1970, with recontoured intake ports and short, round exhausts. They performed well, but casting problems kept them out of production.

Chevelle SS 396 models with the L78 option were potent performers. Photo courtesy of Chevrolet.

tuned big-block engines offered in the early supercars would have nowhere near enough performance to satisfy hard-core street racers of the '60s. They had been weaned on big-port heads, multiple carbs and solid-lifter, rump-rump cams in the super-stock period. Even though the new GTO-class cars could perform as well as the earlier SSs, the street crowd knew there was more to be had with a few extra engine goodies.

The factories were only too willing to answer this demand. In some cases, they had the goodies waiting on the shelves, from earlier NASCAR or drag-race combinations. Usually, the required engineering development and tooling were simple and straightforward. And most important, these optional performance packages could be sold for an extra $300—$400 each. Easy money. Why not?

In the '65—'70 period, practically all the Detroit automakers developed one or more optional performance packages for their standard supercars. They were available by checking off the right items on the order blank. Here is a brief rundown of how the various companies responded to this special demand.

Buick—After their all-new V8 engine was introduced in the 1967 GS, Buick engineers got to work on what became known as the *Stage-I* performance kit. The original package consisted essentially of a high-lift camshaft, high-rate valve-spring/damper assemblies rated at 280 lb at full lift, and lightweight tubular pushrods. Stage-I Buicks also had a 60-psi oil pump, oversize oil-pickup pipe, richer metering for the Quadrajet carb, fast spark-advance curve, high-output fuel pump and chrome valve covers.

This kit was released in late '68. When the 455-CID engine was used in the '70 GS models, the Stage-I kit was upgraded with larger valves in the standard cylinder head—2.12-in. intakes and 1.75-in. exhausts. This added 5—10 HP above 5000 rpm.

It should also be mentioned that Buick developed a very special *Stage-II* engine kit during this period, strictly for dealer installation. This was designed to be installed in a Stage-I car. The Stage-II Kit consisted of a still higher-output cam with higher lift, and new cylinder-head castings with larger intake ports and huge *round* exhaust ports. No special exhaust-manifold castings were made to fit these ports: The owner was instructed to install tubular-steel headers immediately!

Unfortunately Stage-II Buicks never made a mark for themselves because of a flaw in the head patterns or cores. Evidently, the flaw was so bad that it required more than half of the castings to be scrapped. Buick's performance budget didn't permit a redesign of the head, so the whole Stage-II project was dropped. I never had a chance to test one—and never saw a published test. So there!

Chevrolet—Chevy engineers had an easy time coming up with an optional performance package for the SS 396 Chevelle. They just dipped into the Corvette parts bin. The 'Vette was using a 427-CID version of the Mark IV big-block at this time, and practically all parts were interchangeable with the 396 block.

The famous *L78* option package for the 396 consisted essentially of the big-port heads, high-riser aluminum manifold, 780-cfm Holley 4-barrel, the long-duration, solid-lifter street cam, high-rate valve springs, 11:1 compression, forged pistons and crank and four-bolt main caps. It was a very complete street/racing package from any standpoint. It was conservatively rated at 375 HP.

The only flaw was that the large, low-restriction, cast-iron exhaust manifolds from the Corvette wouldn't fit in the Chevelle chassis. So, standard manifolds were used. They probably cost 10—20 HP. But this was still a very strong engine.

By the late '60s, GM front-office restrictions on the horsepower and engine displacement of intermediates didn't have much bite anymore, because of the rapidly increasing weight of the cars. Even the potent L78 Chevelle was somewhat acceptable to GM brass due to its increased weight. Nevertheless, it's been said that Chevrolet's performance boss, Vince Piggins, listed the L78 as a *dealer-*

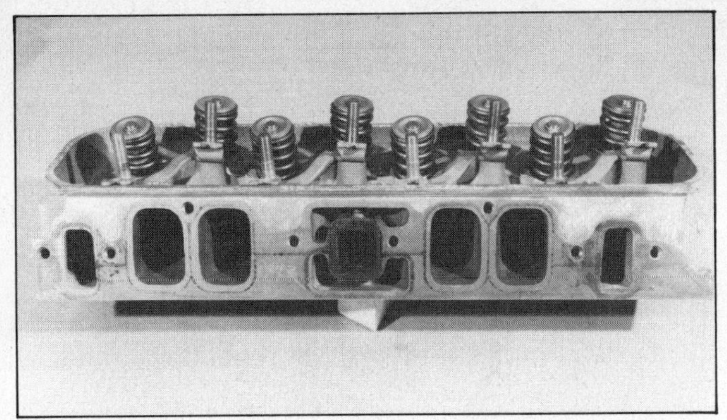

Chevy Mark IV high-performance heads, first seen in 1965, had nearly 4 sq in. of intake-port area. Almost too big for street driving, but fantastic breathing above 5500 rpm. Photo courtesy of Chevrolet.

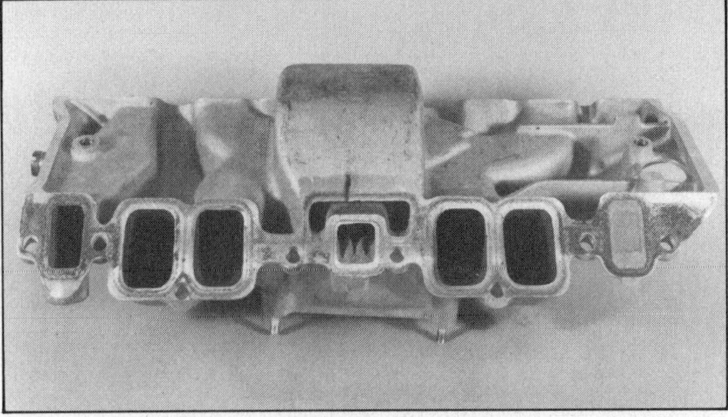

Underside view of Chevy high-riser aluminum intake manifold for the Mark IV performance engines. Manifold was flanged to mount a 780-cfm Holley 4-barrel. Great street-carburetion system. Photo courtesy of Chevrolet.

installed option to ensure its front-office acceptance!

So it was no surprise when Vince decided to up the ante in 1969 and dump the high-performance 4-barrel 427 in a short run of Chevelles, reportedly to make the combination legal for NHRA Super-Stock classes. Called the *L72,* it was essentially the same combination as the L78 except for the displacement. However, the cars came through with 4.11:1 final drive as standard equipment. And the six units that were released with Turbo Hydra-matic-400 transmissions had high-stall torque converters and valve bodies and governors calibrated for 6200-rpm full-throttle upshifts.

These were very quick cars right off the showroom floor. I had a chance to strip-test one of these, with no more tuning than a quick timing check. It turned 13.80s at 103 mph. There seems little doubt that this less-than-100-unit run of L72 '69 Chevelles were Chevrolet's quickest supercars ever.

These '69 427 Chevelles were *considerably* quicker than the fabled *LS6* 454s of 1970. Chevy engineers emasculated the LS6s by forcing them to wheeze through a new low-riser Corvette intake manifold and a dual-snorkel air cleaner. The combination was supposed to reduce noise, but it hurt performance in the process. The 427 L72 combination used the early high-riser manifold and the big AC 14-in. open-element filter. Losing these two items together must have cost 30—40 HP. In fact, CAR LIFE tested an LS6 Chevelle with 3.31:1 final drive and the optional cold-air hood: It could do no better than 14.6 seconds at 99 mph in the quarter mile.

Dodge/Plymouth—Chrysler engineers offered a good dose of image as well as performance with their power kit for the 440-CID Magnum engine. They borrowed the Holley triple 2-barrel carburetion system developed for the 427 Corvette and put it on an Edelbrock high-riser aluminum manifold. Ambient air was routed to the carburetors

Chevy offered their 427-CID high-performance 4-barrel engine in Chevelle, Camaro and Nova in 1969. Rubber gasket around the air cleaner sealed against the hood scoop for fresh-air induction.

In '70, Chevy upped the Chevelle big-block option to 454 CID. But the resulting cars were actually slower than the 427s of the previous year. Photo by David Gooley.

MoPar Six-Packs had the most-effective air scoop in the industry. This is a '69 Plymouth Road Runner. Photo courtesy of Chrysler Corp.

Dodge Division settled on the *Six-Pack* name. This is the hood scoop on a '69 Dodge Super Bee. Photo by Ron Sessions.

Original Six-Pack system in '69 had three Holley 2-barrels on an Edelbrock high-riser aluminum manifold, with vacuum operation of the two end carbs. Chrysler made their own cast-iron manifold in '70. Photo courtesy of Chrysler Corp.

Chrysler's special performance gimmick for the 440 Magnum engine was a neat triple 2-barrel carburetion system. Photo courtesy of Chrysler Corp.

from a very efficient *ram-air* scoop in the special fiberglass hood. The three carburetors gave a total airflow capacity of more than 1000 cfm.

Street flexibility was assured by a unique system of vacuum diaphragms to open the two end carbs as airflow increased through the center carb. Cruising was done on only the center carb. But with the end carbs opening gradually between 2000 and 4000 rpm, there was a very smooth transition from 2-barrel cruise to full 6-barrel breathing. It was almost as smooth as an airflow-controlled 4-barrel.

Chrysler called it the *Six-Pack*. The term still conjures up images today!

But there was more to an optional Six-Pack engine than carburetion. Chrysler wanted to use over 300 lb of valve-spring force to allow shift points up to 6400 rpm. To use

Mopar 440 Six-Pack engines were beefed up with big rods from the Hemi. Extra rod weight required a special crank balancer. But the engines were bulletproof to 6500 rpm. Photo by Mike Crawford.

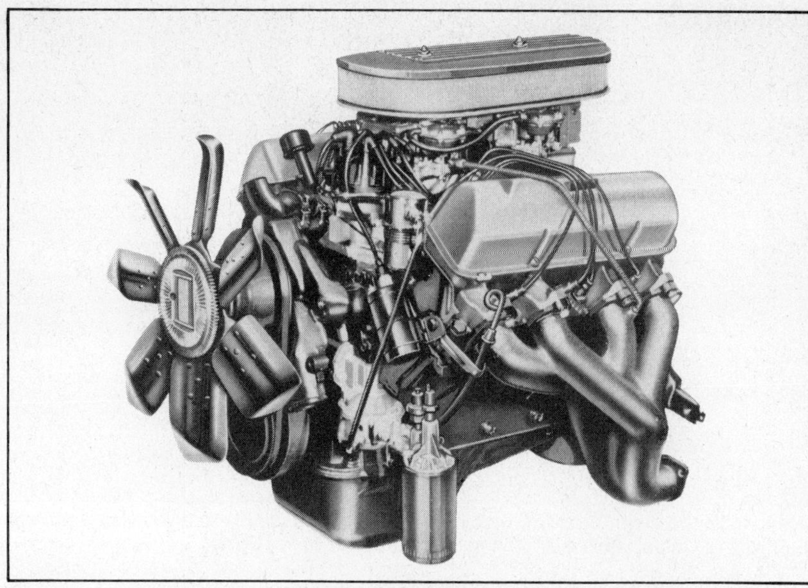

Ford adopted many NASCAR tricks on their 427 street engines in the '65—'67 period—including cross-bolted main caps, side oiling, medium-riser heads, forged pistons and so on. Great performance engine. Photo courtesy of Ford Motor Co.

this much spring without excessive cam-lobe wear, they tapered the lobes in an axial direction. They also used special lifters with a pronounced radius on the bottom surface—which had the effect of reducing unit rubbing loads and producing a slight offset drag to rotate the lifters. Other manufacturers had radiused the lifters before, but not to this extent. It worked! Unlike the street Hemis, Six-Pack 440s suffered no wear problems. I'll discuss the Hemi later.

For additional blow-up insurance, Chrysler used chrome-stem valves, stronger rocker arms, heavy-duty spring retainers, Hemi-type connecting-rod forgings and a 70-psi oil pump. They even carried the beef-up ideas through to the rear axle, using the big 9-3/4-in. Dana unit that was formerly reserved for trucks!

Biggest surprise of all was the price: only $463 for the whole package. It was certainly one of the better performance/durability bargains of the supercar era.

Ford/Mercury—Ford took a different approach to the problem of upgrading standard supercar performance. They often replaced an entire engine rather than modifying an existing one. A good example was replacing the 390 GT engine with the 428 Cobra Jet in '68—'69.

Ford took a similar approach to optional performance. That is, rather than develop a power pack for the 390 when the GT models were introduced, they released a detuned version of the 427-CID NASCAR *medium-riser* engine as a factory option in 1967. You could have the 427 in a GT or most other Fairlane and Comet models by checking the appropriate box on the order blank.

This 427 medium-riser engine was a refinement of the 406 and 427 super-stock combinations of the early '60s. It used what would later become the 428 CJ intake manifold and cylinder-head porting and valves. The 427 medium-riser also had dual 4-barrel carburetion, 11.5:1 compression, solid-lifter cam, forged pistons and crank, cross-bolted mains and low-restriction exhaust manifolds.

Ford offered the 427 8-barrel engine in intermediate Fairlane and Comet bodies in '67. A good combination for both street and drag strip. This '67 Fairlane ran competitively in the SS/B class. Photo by Warren Tanzola.

In short, this '67 427 used all the good stuff Ford had on the shelves at the time. It was a very strong engine with 425 HP. But it suffered from the same malady as all the early super stocks: a cam with too much lift and duration, ultra-high compression and way too much carburetion. The combination was not suitable for smooth, flexible street driving. And these Fairlanes and Comets suffered from high fuel consumption. Only a few brave souls had the nerve to order a 427 for their Fairlane or Comet.

In 1968, Ford tried another tack: Detune the 427 for easier street driving, but retain enough power to be well ahead of the regular 390 GT engine. It was easy. Ford did it by going back to the early 427 heads, with smaller ports and valves, 10.5:1 compression, a milder hydraulic-lifter cam, and single 4-barrel carburetion. The power rating dropped from 425 to 390 HP at 5600 rpm. Street driving

Aside from the diminutive 427 fender badges, there was little to give away a 427-powered Fairlane or Comet. This is a '67 Cyclone. Photo by Steve Christ.

Oldsmobile offered an optional Tri-Carb system on the 4-4-2 in the '66—'67 period. Its progressive mechanical throttle linkage was more responsive than the vacuum setup on GTOs. Air cleaners used open-cell plastic-foam elements.

Many fans don't realize that Olds' W30 Force-Air system was first introduced in 1966. Original design took in air through flexible tubes from the grille area, like this prototype system on a '65 4-4-2. Strictly a dealer-installed option.

Oldsmobile's 1967 4-4-2 was the last to offer optional triple 2-barrel carburetion and the popular tunnel-type rear roofline. The Cutlass body was restyled in '68, and the Quadrajet 4-barrel carb took over. Photo courtesy of Oldsmobile.

was much smoother and more economical. But alas, few liked the *hydraulic* 427 either. Too sluggish. No clatter from solid lifters. No power beyond 6000 rpm. The car promised more than it delivered. Ford missed the mark—again!

It wasn't until Ford brought out the 428 Cobra Jet engine in '69 that they really started to tap the street-racing market.

Many Ford fans don't realize that there was an optional performance package to go with the '69—'70 428 CJ engine: The *Super Cobra Jet* option. It was aimed primarily at serious street and drag-strip racers who needed reliability at high revs. Rather than adding "more" camming and compression, Ford engineers looked at the bottom end. The SCJ package used stronger 427-LeMans rods, an 80-psi oil-pump, a windage tray, deep-sump oil pan and an external oil cooler. Ford threw in a 4.30:1 final drive and locking differential for good measure.

The new SCJ goodies didn't add any power—but you could at least feel safer using the power you already had!

Oldsmobile—Like Pontiac, Oldsmobile decided to use triple-2-barrel carburetion as their major performance option for early 4-4-2s. Why not? It had looks, image, did something for top-end power and was fairly easy to design and tool. Oldsmobile's *Tri-Carb* option was popular on 4-4-2s in the '66—'67 period at an extra cost of around $120. And the system was more responsive than Pontiac's. It used a progressive mechanical throttle linkage rather than Pontiac's jerky vacuum system.

Oldsmobile's next performance package was the mysterious *W30* package. This dealer-installed option consisted of a carburetor airbox with two large flexible Force-Air hoses to bring outside air up from scoops behind the grille, *plus* a wild 308° camshaft and high-rate valve springs to allow 6400-rpm speeds.

The W30 kit really made those early 400-CID 4-4-2 engines turn on. But the extra cost and bother of dealer installation was a tremendous handicap to sales. Only a few-

Olds W30 Force-Air system for the '68—'69 cars had air scoops under the front bumper, feeding through long, 4-in.-OD flexible hoses to a dual-snorkel air-cleaner housing. Photo courtesy of Oldsmobile.

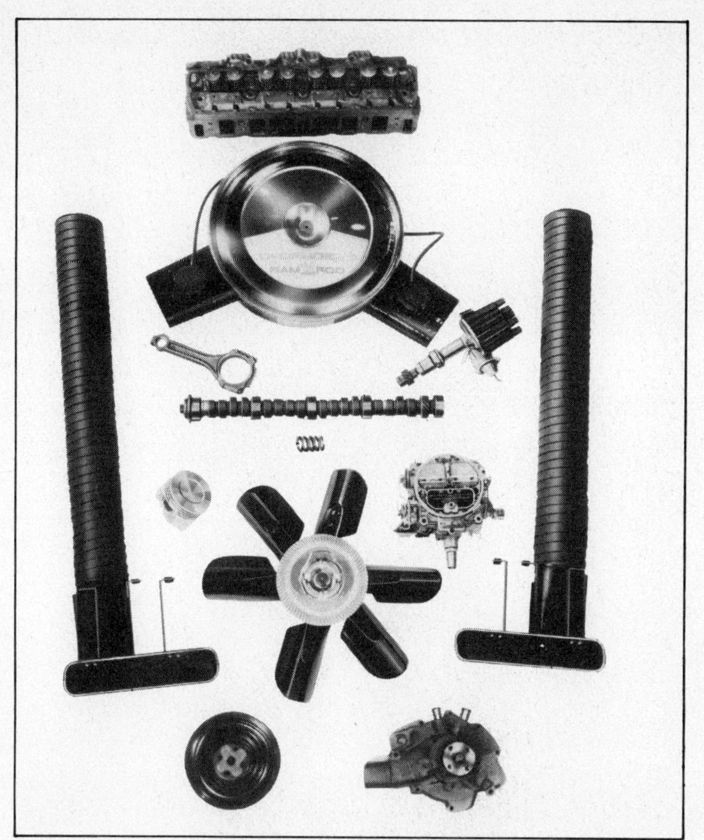

Late Olds W30 package included a lot more than a ram-air system. It included modified cylinder-head castings, special camshaft, valve springs, pistons, water pump, fan, and more. Photo courtesy of Oldsmobile.

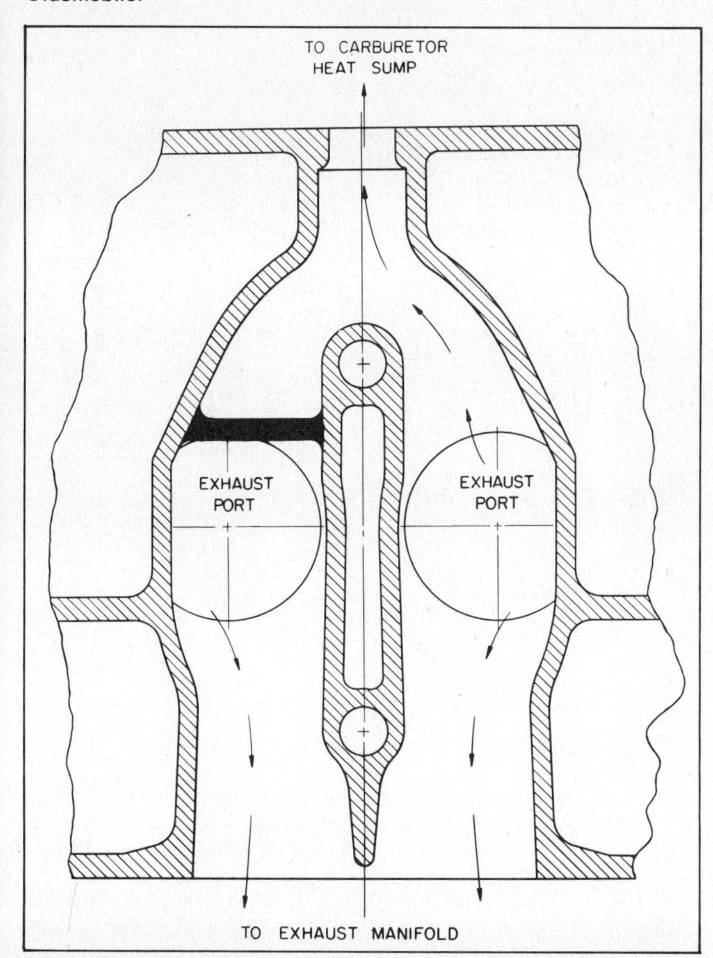

Olds W30 engines of '68—'70 had one center exhaust port blocked to cut off gas flow to the manifold crossover. This improved performance with open exhaust and tubular-steel headers. Drawing courtesy of Oldsmobile.

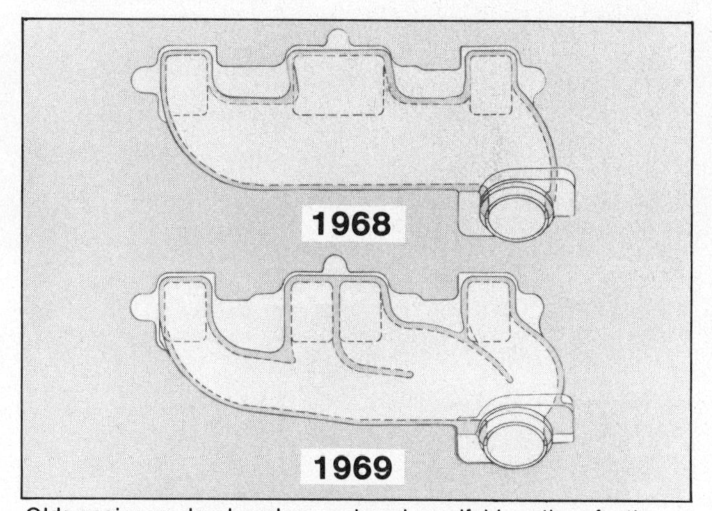

Olds engineers developed new exhaust-manifold castings for the 4-4-2 in 1969. Internal dividers separated pressure pulses for several inches. Split center port added 5 HP when after market headers were used. Drawing courtesy of Oldsmobile.

hundred W30 kits were ever sold. But these were the cars that were winning on strips and streets in the '66—'67 period. Ram air and the long cam duration and overlap complemented each other nicely.

In 1968, the Olds top brass had the courage and foresight to spend a bundle of money tooling up a W30 package for *factory* installation. Considerably more sophisticated than the previous dealer-installed kit, it included special cylinder-head castings with larger intake valves and individual center exhaust ports. Test showed that the divided center ports worked better with tubular-steel aftermarket headers. This was especially noticeable above 5000 rpm.

In '68, Oldsmobile's optional W30 version of the 4-4-2 used a new Force-Air system with scoops under the front bumper. Package included Quadrajet 4-barrel and a very high-lift, long-duration cam. Photo by David Gooley.

Optional Tri-Power carburetion on '64—'66 GTOs used a unique electro-vacuum-operated throttle linkage. But most owners converted to mechanical linkage. Photo by Alex Walordy.

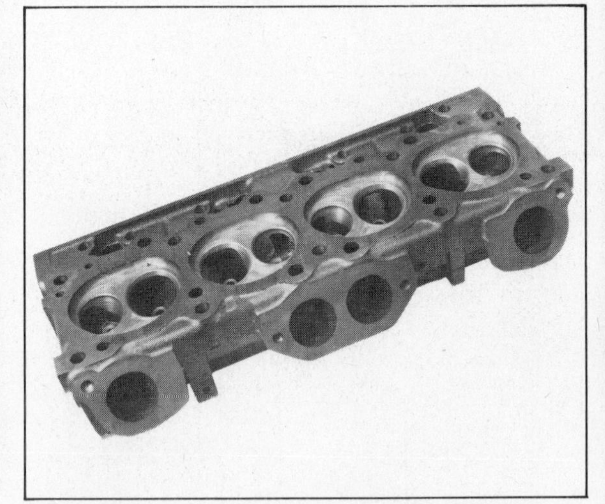

Pontiac Ram Air IV cylinder heads for '69—'70 had round ports and special manifolds. Exhaust flow was not aided, but improved intake porting added 10—15 HP.

Carburetion was standardized with the 750-cfm GM Quadrajet, and for '68, the ram-air tubes pulled from broad, flat scoops under each end of the front bumper. A more-radical 328° cam and higher-rate valve springs increased power in the 5000—6200-rpm range. A declutching fan saved another 10 HP. And to keep it all together, Olds used harder aluminum rod and main bearings, and increased oil pressure. Increased bearing-to-journal clearance was used to get more oil to the journals. Looser piston clearances were used to reduce friction.

With the improved breathing, it's unfortunate that an unrelated characteristic of the '68—'69 400-CID engine worked against all the new power-producing goodies at high rpm. Because Olds wanted to save money by using the 455 crank and rods in the '68—'69 400, stroke was lengthened more than 1/4 in. To maintain 400 cubes, and keep the front-office boys happy, Olds *reduced* the bore size. The extra friction and heavier reciprocating parts were counterproductive. As a result, the long-stroke 400 didn't seem to develop much more power than the '67 short-

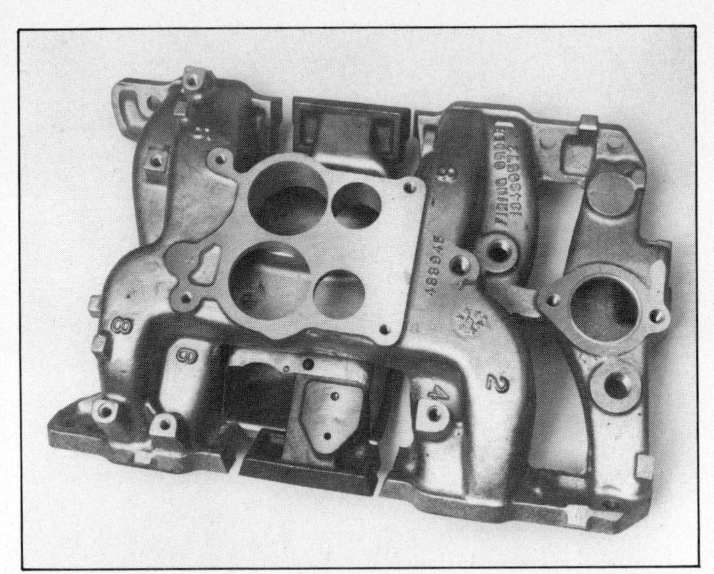

Ram Air IV also had the best intake manifold Pontiac ever made; a beautiful aluminum high-riser, flanged for the 750-cfm Quadrajet. Also used on the 455 H.O. in '71—'72.

stroke engine, despite the improved heads and cam. Oldsmobile fans have been arguing this point for years. But published road and strip tests verify that the '68—'69 W30s were no quicker than the '66—'67 versions.

It wasn't until Olds released the 455 block for the '70 4-4-2 that the W30s really started to show their full potential. Cubic inches saved the day again!

Pontiac—Pontiac's *Tri-Power* carburetion system was already a popular option before the GTO was introduced. So, it's no surprise that it was ordered on a large percentage of Pontiac's new street supercars. In fact, I can't recall seeing a published road test on an early GTO that *didn't* have Tri-Power. This is one reason the standard models earned an early reputation for performance they didn't deserve. The Tri-Power system boosted output at least 10—15 HP.

The first *Ram Air* GTOs appeared in 1966. The package consisted of a functional hood-scoop system, feeding outside air to a foam-sealed airbox around the three carbs. And like the Olds Force-Air kit, the Pontiac option also included a high-output 288° camshaft and high-rate valve springs. The cam and valve gear contributed as much to increased performance as the Ram Air. In fact, the following year, when the three 2-barrels were dropped in favor of a single Rochester Quadrajet, Pontiac adopted an even stronger 301° cam for the Ram Air package. Later that year, they introduced a special cylinder-head casting with large round exhaust ports.

The epitome of GTO engine development came in 1969 with the magnificent *Ram Air IV* engine. This was the most-complete ground-up engineering of a Pontiac performance engine since the Super Duty 421 of the early '60s. The package included big-port cylinder heads with round exhaust ports, aluminum high-riser intake manifold, 750-cfm Quadrajet carburetion, 313° camshaft with 0.520-in. valve lift, 6500-rpm valve gear and special exhaust manifolds. Down below, it had forged pistons and crankshaft, 4-bolt main-bearing caps, windage tray and a high-capacity oil system. It was an impressive package in the catalog, under the hood, and out on the street.

At any rate, many GTO buyers were more than willing to pay $390 extra for it on '69—'70 models. It was one of the industry's most-popular all-out performance options. Clean Ram Air IV examples are extremely valuable.

AGAINST THE STOPWATCH

Below are some more CAR LIFE strip tests of some of these optional engine combinations on GTO-type cars in the late '60s. I have filled in from my own test files where reliable published tests were not available.

Check these figures against those previously shown for the stock GTO-type cars and you can see that these optional performance kits added up to 100 HP over the standard engines. Remember that these kits could be had for a mere $100—$500. Definitely a good horsepower-per-dollar buy.

But there was also a stiff price in driveability and fuel economy. These engines, with their long-duration cams and high-airflow induction systems, were almost as hairy to drive on the street as the super stocks of the early '60s. And even with gas at only 35¢ a gallon, you thought twice about taking off for a Sunday jaunt to a drag strip 100 miles away. You could easily burn 30 gallons for the day with a

Car	Transmission	Axle ratio	Adv. HP	0—30 mph	0—60 mph	1/4-mile e.t. @ mph	Est. Net HP
'66 Olds 4-4-2 W30	M4	3.55:1	360	2.5 sec	6.3 sec	14.8 @ 97	320
'68 Buick GS Stage I	A3	3.91:1	345	3.0	6.1	14.4 @ 93	290
'68 Chevelle SS L78 (396)	M4	3.55:1	375	3.0	6.6	14.8 @ 99	360
'69 Pontiac GTO RA IV	M4	3.55:1	370	2.7	6.2	14.4 @ 98	340
'70 Olds 4-4-2 W30 (455)	A3	3.42:1	370	2.6	5.7	14.4 @ 100	370
From Personal Tests:							
'67 Pontiac GTO Ram Air	M4	3.91:1	360	2.8	6.3	14.7 @ 96	310
'68 Olds 4-4-2 W30 (400)	M4	3.55:1	370	2.9	6.4	14.7 @ 97	320
'69 Chevelle SS L72 (427)	A3	4.11:1	425	2.7	5.4	14.1 @ 101	390
'69 Dodge R/T Six-Pack	A3	3.91:1	390	2.8	5.8	14.2 @ 100	370

Olds got around the GM cubic-inch limit in intermediates by having Hurst install the 455-CID V8 in a select number of 4-4-2s. The GM front office didn't ban Olds from selling the big-engine cars—just banned them from making them! Olds dealers sold the car as the Hurst/Olds. This is a '69 model. Photo courtesy of Hurst Performance, Inc.

THE HURST/OLDS SYNDROME

Legend has it that the first Hurst/Olds was built for none other than George Hurst, president of Hurst Performance, Inc. George was a fancier of the 4-4-2, but thought the concept could be improved. A golden opportunity surfaced when Olds introduced the 455-CID big-block V8 in '68. This potent 380-HP engine was a real stormer, but the GM front office banned engines over 400 CID in intermediates. Olds made an end run around the corporate ban—and arranged for Hurst to drop the 455 V8 into the 4-4-2. Suddenly, Hurst was in the car business.

These cars were by no means strippers. Hurst keyed their marketing campaign to the *executive hot-rod* theme. That first Hurst/Olds was well equipped with Turbo Hydra-matic 400 transmission and 2400-rpm high-stall torque converter, Hurst Dual/Gate shifter, Anti-Spin rear axle, low-profile Goodyear Polyglas F60 X 15 tires on 7-in.-wide mag-style wheels, and front disc brakes.

In the go department, the big Olds was no slouch with a dual-inlet Force-Air induction system, high-overlap camshaft and modified spark-advance curve. A Rallye Suspension package improved the already impressive 4-4-2 handling with even higher-rate springs, firmer shocks, bigger anti-sway bars and harder suspension bushings.

With the standard 3.42:1 final-drive ratio, the 455 V8 was loafing at highway speeds, with plenty of reserve for passing. The engine had gobs of low-end torque, 500 ft-lb to be exact, for good low-speed response. Olds even encouraged buyers to choose optional air conditioning—something not even available on most supercars of the era.

In '69, Hurst added a rear spoiler. The spoiler was stanchion-mounted on the trunk lid. Besides oozing image, the spoiler was supposed to decrease drag and improve stability at high speeds.

All totaled, Hurst built 1421 modified Olds 4-4-2s in two years. But in 1970, GM lifted its 400-CID limit on intermediates and Olds decided to make its own 455-CID 4-4-2s. Olds figured they could make them in-house, quicker and cheaper. Hurst was out of a job.

There was no Hurst/Olds in '70 or '71, but in '72 Hurst got the nod to build 629 Indy 500 Pace Cars. Equipment was similar to previous models.

Hurst and Olds teamed up to build more than 5000 Hurst/Olds in the next three years. But by the mid-'70s, the emphasis on performance was reduced. It was replaced by special features such as an electric sunroof, padded-vinyl landau roof, digital-readout speedometer and tach, swivel bucket seats, a security alarm system and tamper-proof wheel locks. Instead of the Hurst/Olds being based on the 4-4-2, Hurst was modifying Cutlass S coupes.

In '75, Hurst introduced the first *Hurst Hatch*—twin removable roof panels. The option proved so popular that Olds green-lighted Hurst to build garden-variety Cutlasses with the open-air roof option—sans the other Hurst/Olds performance goodies. By this time, Hurst was offering an "economy" 350-CID V8 in the car as well. It was indicative of the times.

Hurst pitched Olds again in '77 and '79, but Olds said no dice. The two would not collaborate on a performance model again until the performance renaissance of the '80s.

few runs down the strip!

So with all the brilliant performance engineering that came out of Detroit in the supercar era, you still had to face the age-old compromise between acceleration, driveability and fuel economy. Nothing had really changed.

AN OVERVIEW OF THE GTO-TYPE CARS

I want to emphasize once more that the GTO-type cars of the late '60s were an answer to the shortcomings of the super stocks of the early '60s. The super stocks were almost too highly tuned for street driving, resulting in

By 1975, the Hurst/Olds had taken on posh and lost some punch. This model had the padded-vinyl roof and *Hurst Hatch* options. Photo courtesy of Hurst Performance, Inc.

Legend has it that the first Hurst/Olds was built for George Hurst. This '68 model was touted as "the gentleman's hot rod." Photo courtesy of Hurst Performance, Inc.

noisy, rough running, hard starting, balky cold-weather operation and terrible fuel economy. The GTO-type cars were a turnaround. By using a mildly tuned big-block, V8 in a lighter intermediate-size car, automakers were able to get almost the same power-to-weight ratio—and thus equivalent acceleration—while still retaining smoothness and fuel economy. It was as simple as that.

And for that small percentage of buyers who insisted on even more performance than the super stocks gave, it was a simple matter to install optional power packs that would give up to 100 more horsepower for a few-hundred dollars extra. These were subject to some of the same hairy driving characteristics as the early super stocks—*but not quite so bad.*

Which brings us to a look at the behind-the-scene engineering that made it possible to get this remarkable street/highway performance *without* a big sacrifice in driveability, ride quality and handling.

In '74, Hurst got the nod to pace the Indianapolis 500. These are valuable collector cars today. Photo courtesy of Hurst Performance, Inc.

Beneath the skin of the 4-4-2 was plenty of muscle. Photo courtesy of Oldsmobile.

Engineering for Performance

Behind the drawing board

HOT RODDING BECOMES A SCIENCE

In the early days of the automobile—and even into the '50s—the idea of *hopping-up* the performance of a production car was not considered a part of legitimate automotive engineering. This was a business for the untrained backyard hot rodder. When the car companies did occasionally get into racing in one way or another, as often as not they would call in an *untrained* hot rodder to supply the special parts they needed. There wasn't much performance know-how around Detroit in the early days.

All this changed when mass-produced high-performance cars became big business in the 1960s. With multimillion-dollar profits at stake, it didn't take company executives long to set up special departments. They were manned by dozens of full-time engineers doing the very special design and testing work these cars required. Hot rodding became respectable almost overnight.

I must emphasize that this new generation of Detroit performance engineers didn't just copy and refine what backyard hot rodders had been doing for years. They plowed new ground. They dreamed up new ideas. They developed new equipment that completely revolutionized the age-old compromise between brute speed and acceleration and practical street driveability. In these areas, the college-trained engineers proved more than equal to the intuitive designers who came up through the ranks via the hot-rodding sport.

Here are a few of things the engineers did in the period between 1965 and 1970 . . . .

Above: Holley 4150 4-barrel carbs with center-hung floats were very popular on high-performance cars of the late '60s. Center inlet minimized fuel starvation during cornering. This is the 735-cfm size on the Ford 428 Cobra Jet engine. Photo by Thomas E. Stuck.

CARBURETION

When the 4-barrel carburetor was first developed in the early '50s, performance fans thought it was the ultimate answer for big increases in top-end power. But if you were to measure the airflow capacity of one of those early 4-barrels, it would be only 400—450 cfm—barely enough for a 200-HP engine. True high-performance engines in those early days required three 2-barrel carbs or two 4-barrels.

Two developments in the mid-'60s gave the single 4-barrel a new lease on life. One was Rochester's *Quadrajet* design, using two very small primary venturis and two very large secondaries. A spring-loaded air valve for the rear barrels prevented over-carburetion at low speeds by opening gradually between about 2000 and 4000 rpm. Thus, at wide-open throttle (WOT), it had a total airflow capacity of around 700 or 750 cfm with all four barrels open. But for cruising at low and medium speeds, fuel was fed only through the small front venturis to maintain airflow velocity and give optimum smoothness, fuel economy and throttle response. By merely adjusting the jetting and secondary air-valve spring, GM engineers could use the Quadrajet on engines displacing 300—500 CID.

Rochester Quadrajet had small primary venturis for good fuel economy—large secondaries for performance.

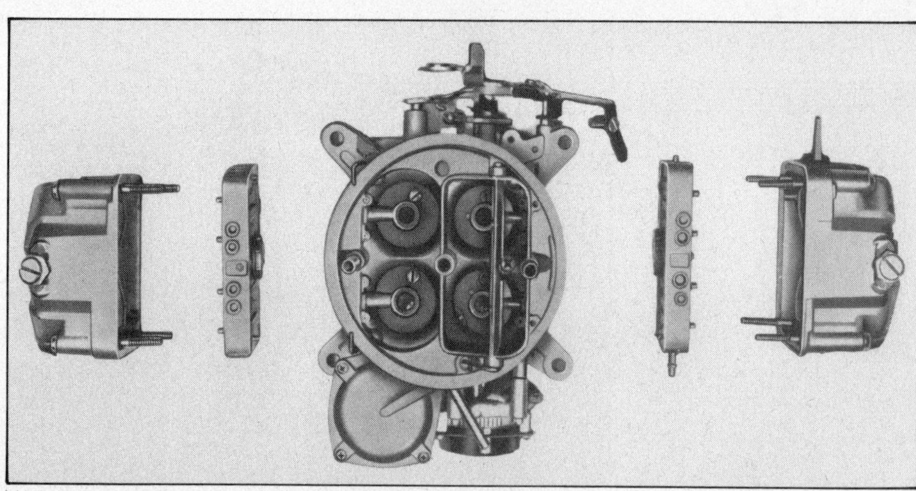

Jet changes were easy with Holley 4-barrel carbs because of the quick-off fuel bowls and metering blocks. Same goes for Holley 2-barrels used on triple-carb manifolds.

Holley engineers achieved all this, and more, by updating their 4150 4-barrel for high-performance use in 1965. A 780-cfm version was used on the '65 Corvette 396. Like the Quadrajet, its secondary throttles were controlled automatically by airflow through the *primary* barrels. A vacuum diaphragm opened the secondaries gradually as engine speed increased. Further, all the metering jets were put in two metering blocks that could be readily removed to fine tune the jetting. Float bowls were designed with narrow width and were center hung so air/fuel mixtures wouldn't be greatly affected by forces encountered during hard cornering. Fuel surge had always been a problem with earlier 4-barrel carbs on high-performance cars. The new Holley 4150 was smooth and stable under almost all conditions.

Holley engineers developed another unique carburetion system that combined image, flexibility and a total flow capacity of more than 1000 cfm. They used three big 2-barrel carbs mounted in-line. The two end units were opened automatically by vacuum diaphragms, according to airflow through the center carb. The gas pedal controlled only the center carb. Actually, this three-carb setup used the same principle as Holley 4-barrels with vacuum-secondary control. The system proved extremely smooth and flexible in everyday street driving. It had reasonable fuel economy while cruising on the center carb, yet the three-2s had tremendous flow capacity when all six barrels opened up in the 3000—4000-rpm range. The system was used on 427 Corvettes, and on both of Chrysler's Six-Pack V8s of the '69—'71 period.

AIR FILTRATION

Auto engineers never thought much about air-cleaner restriction as a performance-limiter up to the early '60s. They were more than satisfied with the tiny dry-paper and wire-mesh filters and dual-snorkel silencers on the early super stocks.

Chevrolet designers introduced the first true "400-HP" air cleaner on the big-block Corvette in 1965. It was used in conjunction with the new 780-cfm Holley high-performance 4-barrel. The filter was pleated paper with a diameter of 14 in. and height of 3 in.—more than 130 sq in. of pro-

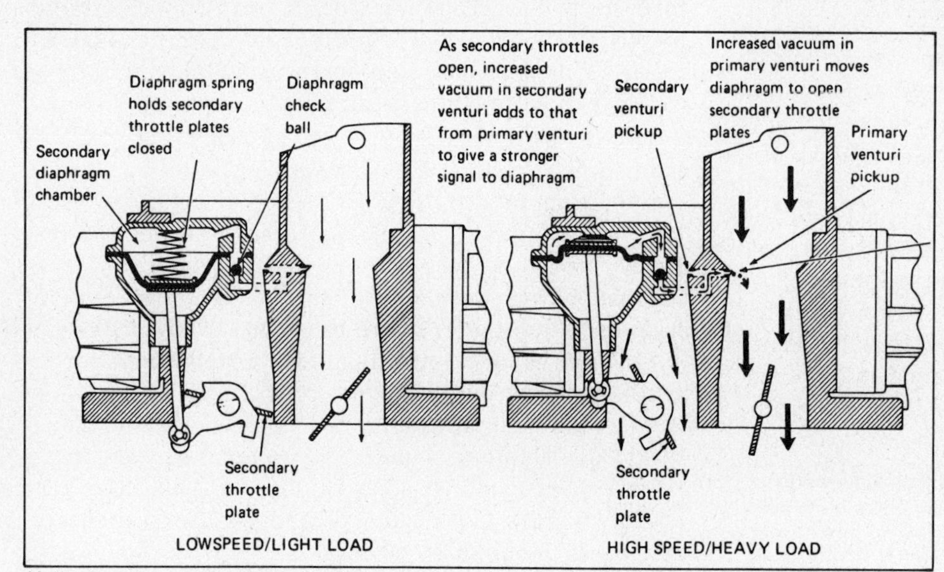

Holley's unique secondary-throttle control used a vacuum diaphragm to open the rear barrels gradually as airflow increased through the fronts. Very flexible on the street.

In the late '50s and early '60s, Corvette 4-barrel engines used a pleated-paper filter in a louvered housing.

By the late '60s, Chevy used a much-bigger 14 X 3-in. paper filter with no housing around it. Airflow was improved over previous designs.

Chrysler used a 14 X 2-in. paper filter in a large round housing for the two 4-barrel carbs on the Street Hemi. The housing gave some silencing with little restriction. Photo courtesy of Chrysler Corp.

jected area. An expanded-metal screen supported the element around its full circumference. Airflow restriction was practically zero, even at 6000 rpm. Subsequent tests showed this filter was worth as much as 15—20 HP over a conventional dual-snorkel closed design on a good-breathing engine like the big-block Chevy. Yet it was very efficient in filtering the air.

This same 14 X 3-in. paper filter was used on several later Chevy engines. Its performance was never surpassed by other filter designs. Other companies hated to copy it directly, so they applied its basic principles to other filter shapes and sizes.

The principles were simple: lots of filter-face area, no restrictive silencer enclosing the filter and a fairly coarse filtering media. This resulted in big, space-consuming, noisy air cleaners and filter elements that had to be changed fairly often. But breathing restrictions were minimal. As for induction noise, the power roar of an open-element air cleaner at wide-open throttle quickly became part of the '60's street-scene music.

THE RAM-AIR PHENOMENON

The idea of ramming cool outside air to the induction system, rather than drawing in hot underhood air, is as old as the automobile itself. Some of the super-stock race cars of the early '60s had efficient hood scoops. The 427 Ford Thunderbolts of 1964 used large flexible hoses to duct air from the headlight openings to an airbox atop the two 4-barrel carbs. It was very effective, both in collecting outside air and in promoting a slight ram effect at speed.

Pontiac was the first company to offer an optional, factory-installed *Ram Air* system. As installed on the 1966 GTO, it was a strong image feature. The package was impressive with neatly styled air scoops on the hood, dumping into an airbox that enclosed the three 2-barrel carbs. The airbox was sealed when the hood closed against a urethane-foam gasket. It was a sure-fire dynamite attention-getter when you popped the hood at a drive-in.

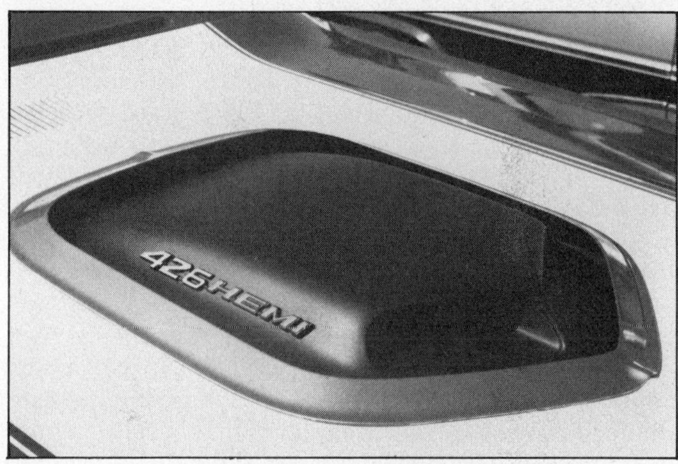

Shaker hood scoops were popular around 1970. The scoop was mounted on the air cleaner and carb, projecting through the hood—so it shook and jiggled as the engine moved on in its rubber mounts. This one's on a '70 'Cuda. Photo by Mike Crawford.

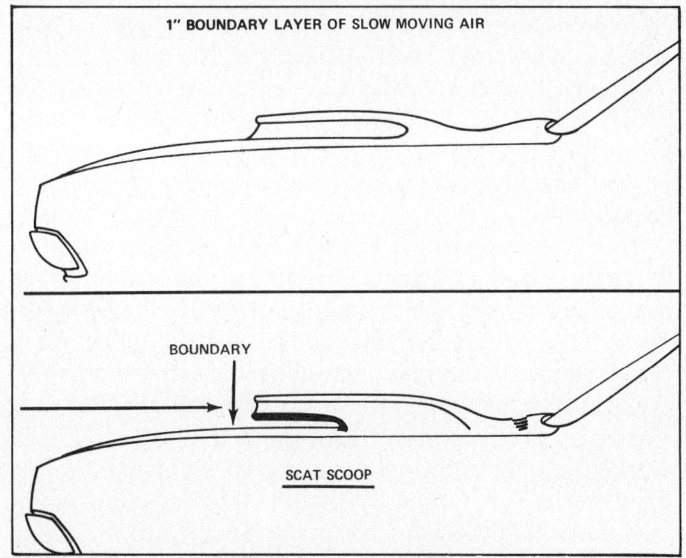

Chrysler engineers developed the most-efficient hood scoop of all around 1970. The scoop opening was raised above the hood surface to avoid the boundary layer. Good for 10—15 extra HP above 80 mph. Dodge used it on the '70 Challenger T/A. Photo courtesy of Chrysler Corp.

Chevrolet's *Cowl Induction* drew air from the high-pressure area at the base of the windshield. This is the system on a Z/28 Camaro. Photo by Dick Wilkins.

1969 Z/28 Cowl Induction system used a large rubber housing around the air filter. It sealed against the hood when closed, to ensure a supply of cooler forced air to the engine.

To back up the image the scoops promised, Pontiac threw in a high-performance cam and high-rate valve springs with the Ram Air package. The cam and springs did improve top-end performance, but not because of the air scoops. As a matter of fact, these hood scoops were very *inefficient.* Pontiac's hood scoops were so low that the openings were mostly in the *boundary layer*—stagnant air next to the hood surface. Regardless of the names, there was no ram pressure with this location, only cool-air feed.

In mid-1966, Oldsmobile offered a dealer-installed *Force-Air* system. Olds engineers took a different approach to ram air. They located the air inlets directly behind the grille—a high-pressure area. Two large flexible tubes were routed to an airbox that enclosed the carbs. This layout was effective in supplying some ram pressure.

When Oldsmobile went to under-bumper inlets in 1968, the ram effect was even more pronounced. Tests showed they added up to 10-HP more at 100 mph than at 50 mph. Ram pressure increases as the *square* of car speed. Theoretically, it's worth a 1—2%-HP boost at 100 mph.

When Chrysler set out to design a ram-air system for their '70 340 *Six-Pack* cars, they decided to use a hood scoop for induction air. Chrysler engineers were more concerned with function than cosmetics, so they went about it scientifically.

Tests were conducted at the Lockheed wind tunnel in Marietta, Georgia. Results showed that the bottom of the scoop opening had to be at least 1.5-in. above the hood surface. This minimized the boundary-layer effect, and provided maximum ram pressure to the carburetors. This is the reason for those huge raised scoops on the fiberglass hoods of Six-Pack cars. They looked hairy—and they worked. Dodge went with the special scoop for the '70 Challenger T/A, while Plymouth opted for the more-traditional NACA flush scoop in the AAR 'Cuda.

Chevrolet used still another ram-air idea in the '68—'70 period. They took advantage of the high-pressure area at the base of a moving car's windshield. Some Chevrolet systems used a reverse scoop on the hood to carry air *forward*

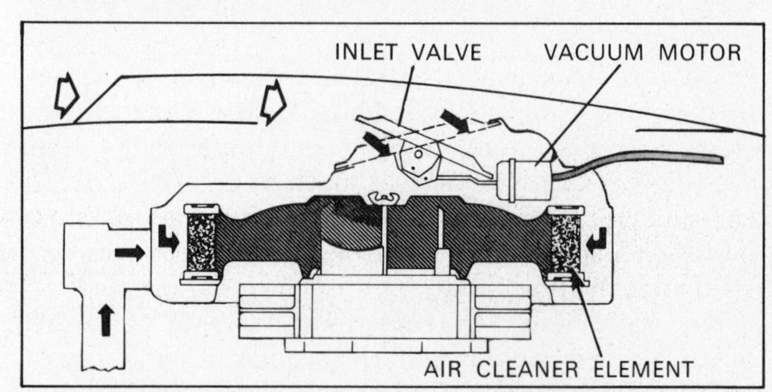

By the late '60s, most ram-air systems had some means for shutting off cold air when cruising—to allow leaner carburetion and better fuel economy. This was the Ford system for Mustangs and Torinos.

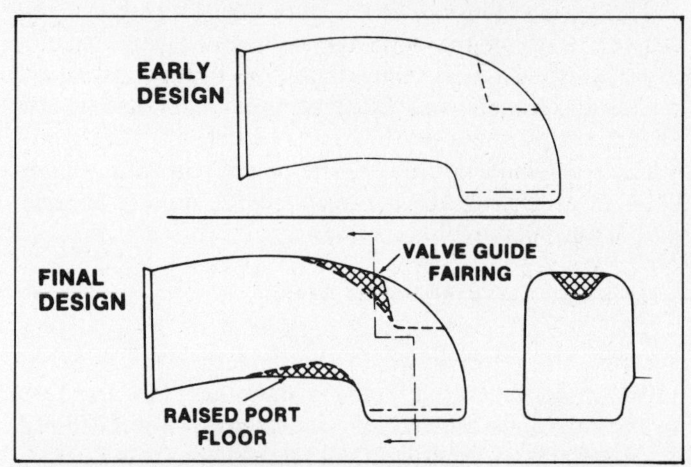

Cylinder-head-port design progressed rapidly in the mid-'60s, as some companies developed airflow test rigs. This shows how intake ports were given a venturi shape to improve flow at moderate valve lift.

from the base of the windshield to the airbox. Remember *Cowl Induction?*

Other Chevrolet systems routed cold outside air from the plenum chamber that feeds the car's ventilation/heating system. Chevrolet pioneered the use of cold-air feed in their '63 Z11 NASCAR package. Tests conducted at the Daytona Speedway in August, '62 indicated a 60F drop in intake-air temperature with the system. It resulted in a 5-6% HP increase. There was definitely a slight positive pressure available here for cool carburetor air—though not as much as with a forward-facing scoop.

The last degree of sophistication for these ram-air systems came in 1970. In cold or rainy weather, ram air could actually hurt engine performance and driveability. The solution was to provide a manual- or vacuum-operated door to shut off outside air when not needed. Several of the cars offered this much-appreciated refinement. But that was the last gasp. Ram air disappeared a year or two later, along with the supercars that spawned it.

CAMS AND VALVE GEAR

Most super-stock engines of the early '60s that used mechanical valve lifters could turn 6000—6500 rpm with no trouble. But mechanical lifters were definitely a no-no on the new breed of GTO-type cars introduced in the mid-'60s. The problem wasn't so much lifter noise. It was the frequent service required to maintain valve clearance. Incorrect valve clearance meant lost power, increased noise and possible engine damage in extreme cases.

Detroit sales people knew full well that their youth/performance market—the young guys buying GTOs and 4-4-2s and SSs—would never observe the strict tuneup schedules as required of the early super stocks. They'd run these cars into the ground, with little more periodic maintenance than oil changes. When 50,000 a year of *any* type of car were sold, it was assumed that a major portion of these cars wouldn't be maintained beyond minimum standards. It was an entirely different deal than when Detroit was selling only 1500 super stocks a year.

So, they decided hydraulic lifters were a must. The problem then became one of providing a high-enough rev range before the lifters pumped up, floated the valves and shut the engine off.

One effective weapon was stiffer, high-rate valve springs. The limit was increased friction between the cam lobes and lifters. Overly stiff springs can cause premature lobe/lifter wear. This wear will kill an engine's performance practically overnight by reducing valve lift and duration. It will also send metal particles through the engine's lubrication system.

We know now that the practical limit for everyday street driving is around 300 lb of spring force at full valve lift. But a number of supercars in the late '60s had 320—330 lb of force at full lift. Cam-wear problems were rampant. Owners would be scratching their heads after only a couple of thousand street miles. It wasn't all kicks, guys!

Another problem was that valve-spring force would gradually drop off due to fatigue. Valve float could occur 400 rpm sooner after 2000 miles. And sometimes, the springs would simply break. Big-block Chevys were notorious for this. Oldsmobile engineers solved spring-breakage problems in 1966 with a special *heat-set* treatment: The springs were heat-treated under stress, so force would remain stable at normal temperatures. Chrome-silicon- and vanadium-steel alloys also helped to increase spring strength and fatigue resistance.

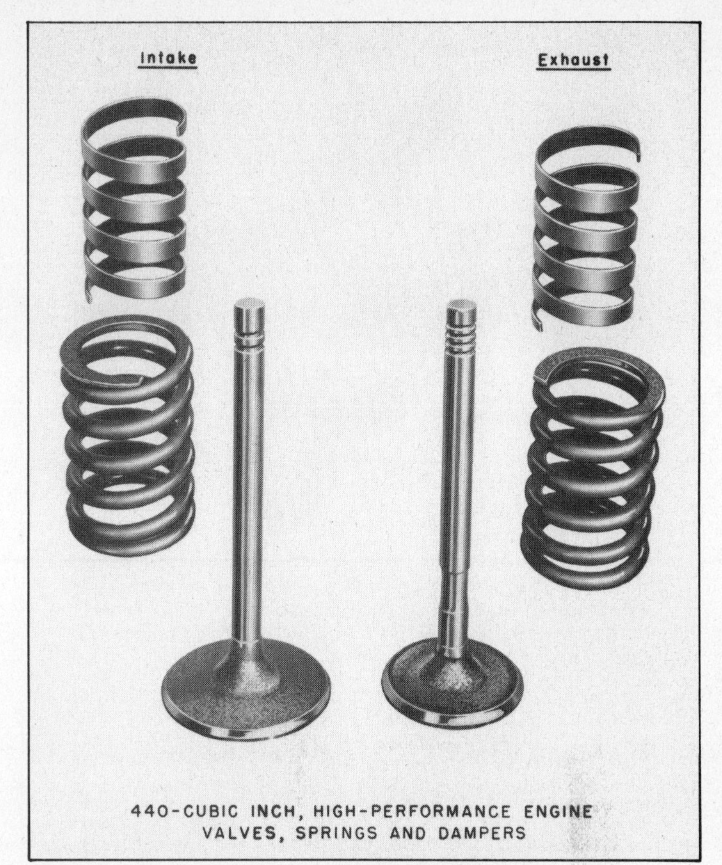

Much development went into valve trains to allow 6000+ rpm speeds with hydraulic lifters. Many engines used damper coils inside the valve springs. Photo courtesy of Chrysler Corp.

Some companies also looked at hydraulic-lifter design. On some engines using stud-mounted rocker arms—Chevy, Pontiac, small-block Ford—the rocker arm could be adjusted to simulate solid lifters. How you say? The trick was to put a heavy-duty retaining ring in the lifter body to prevent breakage and adjust the rocker-arm nut so the lifter plunger was near the top of its travel. This was not a factory adjustment, but racers were urged to do it.

On engines with shaft-mounted rockers—big-block Ford, and B-Block MoPar—this adjustment wasn't possible. For these engines, lifters were designed with extra *leakdown* clearance around the plunger. The result was less tendency to pump up and prevent valve closure at high rpm.

Eventually, Detroit engineers developed hydraulic-lifter engines that could turn at least 6200 rpm with reasonable camshaft and valve-spring life. Compare that with a limit of 4000—4500 rpm for hydraulic lifters in the early '50s.

BOTTOM-END BEEF-UP

Some super stocks of the early '60s were sold without any warranty against engine or drive-line breakage. This wasn't possible with the mass-produced GTO-type cars of the mid-'60s. Result: The engineers had to analyze bottom-end stresses in normal street operation, followed by selective beefing up to assure reliability and durability. This work was especially critical because of the need to count pennies. With the number of cars going out the door, the

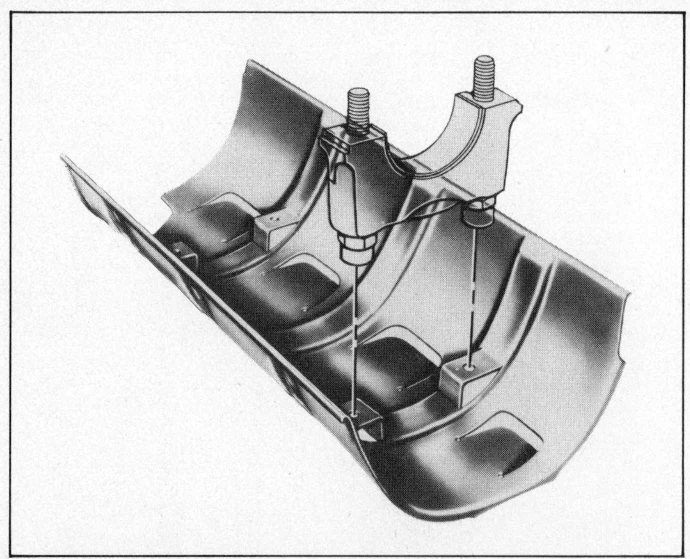

Crankcase windage trays were widely used in the supercar era to reduce viscous drag on the crankshaft and rods. This Chrysler design saved 12 HP at 6000 rpm in the 340 V8. Photo courtesy of Chrysler Corp.

Plymouth Road Runner and Dodge Super Bee used small, low-restriction mufflers and big-diameter pipes. Chrysler was always a leader in efficient exhaust systems on cars big and small. Photo by Phil Chisholm.

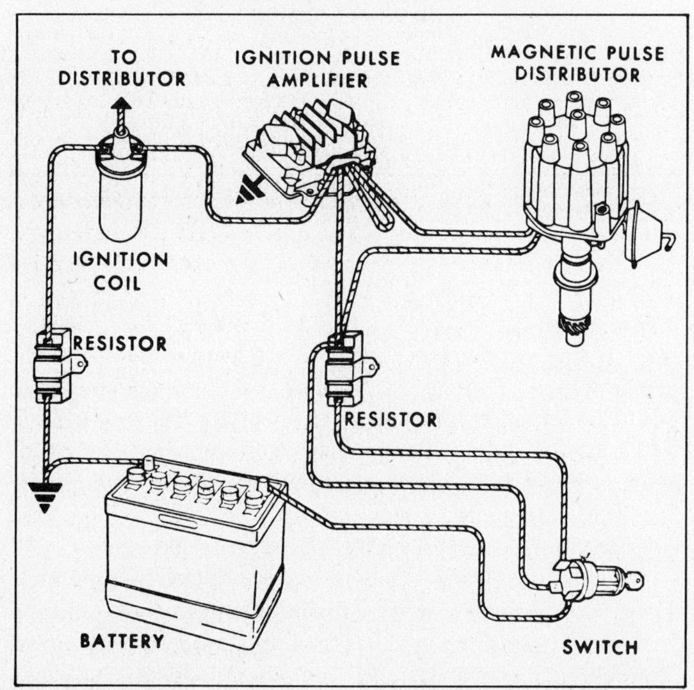

GM's Delco Division introduced a breakerless electronic-ignition system in 1963, using transistors and a magnetic-trigger system. A great improvement over breaker systems. It was optional on some Chevy models in '66—'67. Drawing courtesy of General Motors.

Detroit bean-counters weren't going to permit using goodies that weren't absolutely needed.

Generally, the standard supercar engines, with a rev range of perhaps 5500—5800 rpm, could get by with minimal attention. A typical beef-up might include a set of harder copper-lead or aluminum main and rod bearings, a 60-psi oil-pump, 1/16-in.-larger rod bolts, and so on. Nothing radical.

But when the companies offered engines that could wind to 6500 rpm—maybe with extra carburetion, high-performance camming, high-rate valve springs, and such—they also made major changes to the bottom ends. Such modifications included forged-steel crankshafts, forged pistons, beefier connecting-rod forgings, reinforced cylinder-block castings and four-bolt main-bearing caps. To aid lubrication, enlarged oil passages, extra oil-sump capacity and high-volume oil pumps were often used.

Bottom-end stresses increase as the *square* of rpm. So these loads are almost half again as high at 6500 rpm as at 5500. Needless to say, the engineers had their work cut out for them. It was especially difficult work when you consider that the companies had to stand behind the hardware with at least a 12-month/12,000-mile warranty. Some manufacturers offered a 5-year/50,000-mile warranty.

A neat trick found in some high-performance engines, was the use of the *windage tray.* As previously discussed, the tray is installed directly underneath the crankshaft. It reduces oil whipping by the crank, which causes excessive splashing, aeration and power loss. The simple $5 sheet-metal plate proved to be worth an extra 10—12 HP above 5000 rpm.

IGNITION

Highly tuned street engines are always a challenge to the ignition man because rich air/fuel mixtures often foul plugs in stop-and-go driving. High-voltage spark helped control this nuisance.

Detroit's supercars were important to ignition evolution. They pioneered electronic ignition a decade or so before it was used on standard cars. Ford offered the Autolite transistor ignition as optional equipment on some high-output engines in the mid-'60s. With this system, the majority of the coil's primary current went through the transistor, bypassing the breaker points. With point arcing removed as a durability problem, a much-higher primary current could be used. This in turn gave 30—40% more secondary voltage to fire the plugs—doubling the service life of breaker points.

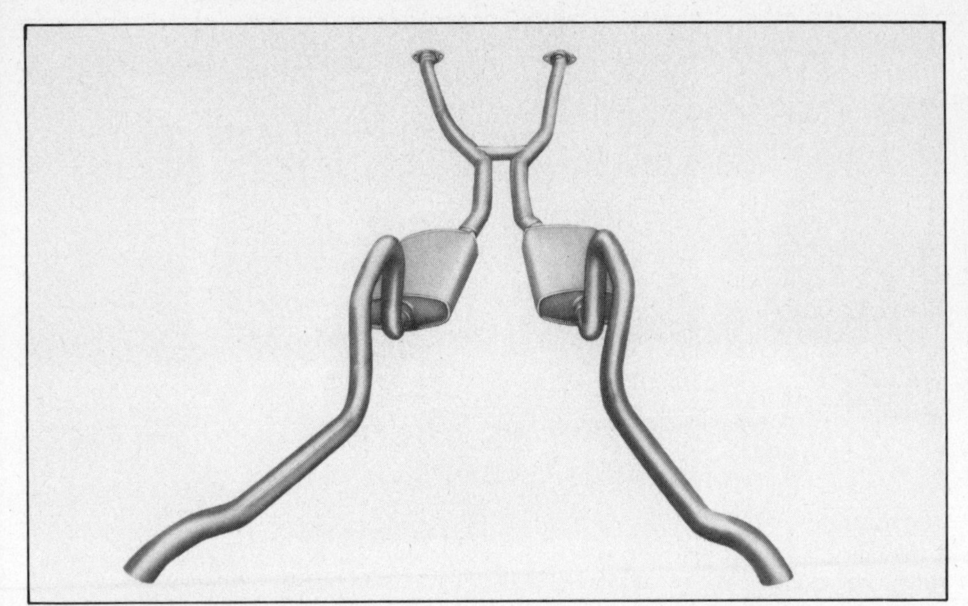

MoPar 440s and Street Hemis used 2-1/2-in. exhaust pipe, big free-flow mufflers, and a unique crossover at the headpipes. The crossover was said to add over 5 HP. Photo courtesy of Chrysler Corp.

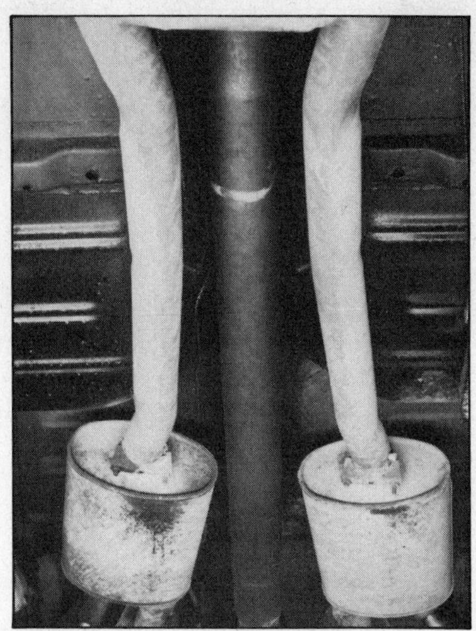

This enterprising Camaro owner replaced the poor factory exhaust system with custom 2-1/2-in. pipes and short mufflers from a turbocharged Corvair. He lost ground clearance, but gained 40 HP. Photo by Phil Chisholm.

GM's *Delcotronic* system was offered as standard or optional equipment on several Chevy, Olds and Pontiac performance engines in the late '60s. Identified by a unique red distributor cap, the system used magnetic-pulse triggering instead of points. A capacitive-discharge booster circuit gave a high-voltage secondary wallop that could fire a plug that was completely coated with carbon. This system was doubly effective on the street because it eliminated contact-point pitting and gradual loss of spark timing because of rubbing-block wear.

GM's Delcotronic was really a great ignition system. But only a few-thousand cars ever were fitted with it. Perhaps the option's $100+ price tag scared 'em off.

EXHAUST SYSTEMS

In the '50s, engineers were well aware that V8 performance could be enhanced by splitting the exhaust flow into dual systems. In fact, there's a primary law of gas dynamics that says *flow restriction,* or back-pressure, in a given channel increases as the square of velocity. Theoretically, using two channels cuts the velocity in half—and reduces restriction by 75%.

It wasn't until the '60s that engineers began to realize there was a lot more to exhaust efficiency than dual exhausts. Super stocks of the early '60s pioneered streamlined and split-flow exhaust-manifold castings. These became a fixture on practically all later supercar engines.

The effect of exhaust-pipe *diameter* came under scrutiny in the mid-'60s. It follows from the above gas-flow law that velocity is inversely proportional to the square of pipe diameter. Researchers found that an increase from, say, 2 to 2.5 in. was almost like adding another pipe to the system. The bigger pipe cost more, weighed more and required more space to bend around chassis obstructions. But it proved a must to gain the necessary breathing for 400-CID-and-larger engines. Where a 1-3/4-in. pipe was popular in the '50s, 2-, 2-1/4- and 2-1/2-in. pipe was prevalent in the '60s.

Muffler design has always been a compromise between noise and restriction. But in the '60s, it became apparent that you could stop high-frequency sounds with less restriction than it took to control low frequencies. This led to sophisticated *low-restriction* muffler designs that muffled annoying high-frequency sounds nicely—but allowed the lows to come through. Low-frequency sounds were the characteristic burbling sounds that became the trademark of a high-performance car. In fact, the sound was literally *tuned* by Detroit marketing types to what they thought the drive-in crowd wanted to hear! And all the while, restriction was held to a minimum.

Another interesting supercar exhaust-system development was introduced by Chrysler. They used a simple crossover pipe to connect the two headpipes ahead of the mufflers. This damped the high peaks of the pressure waves and actually reduced noise and restriction at the same time by doubling muffler capacity. Chrysler produced some of the most-efficient exhaust systems of the supercar era.

CLUTCHES AND MANUAL TRANSMISSIONS

Needless to say, clutches took quite a beating on street supercars. Design efforts were concentrated mostly on using maximum possible spring force on the pressure plate, without substantially increasing pedal effort for street driving. Results were spotty. Excessive clutch-pedal effort was a frequent criticism of magazine testers. Take it from me, those high-effort super-stock clutches were brutes to operate.

Another alternative was to increase clutch diameter, so torque capacity could be increased without the use of ex-

Borg & Beck double-disc clutch solved the torque-capacity/effort dilemma. It made it into production on some '70-model Chevys and Oldsmobiles, but got canned due to high cost.

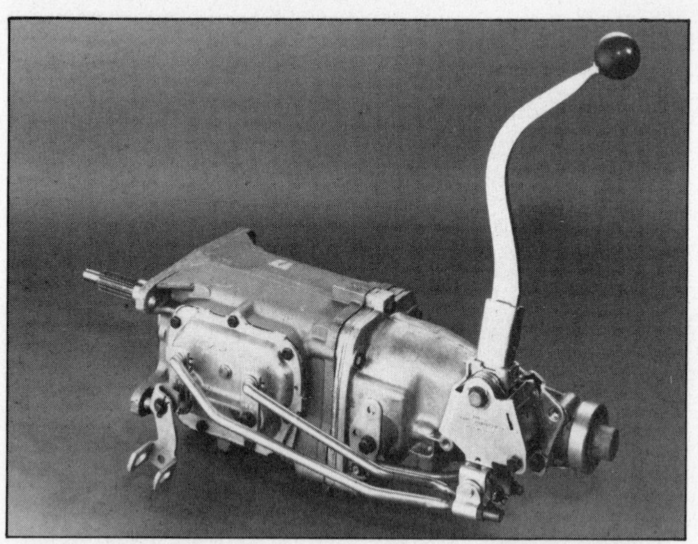

In the late '60s, GM's aluminum-case Muncie M-21 four-speed was available weighing less than 70 lb. A variety of gear ratios was offered, and some GM divisions used a Hurst floor shifter. Photo courtesy of Hurst Performance, Inc.

cessive spring force. The bug here was increased clutch-disc inertia. You couldn't make fast shifts without clashing transmission synchros. The final effort was to upgrade the friction-facing material to take more heat without fading. This helped.

Borg & Beck engineers solved the torque-capacity/effort problem with a unique *double-disc* clutch. By doubling the potential friction area, they were able to get high torque capacity with low spring force, low pedal effort, and with a relatively small overall diameter of 10 in. It made an excellent high-performance street clutch and was used as factory equipment in 1970 on some Corvette and Olds W30 models. But high cost discouraged extensive use of this clutch—especially as automatic transmissions were becoming popular. The Borg & Beck double-disc clutch disappeared from the option list after a couple of years.

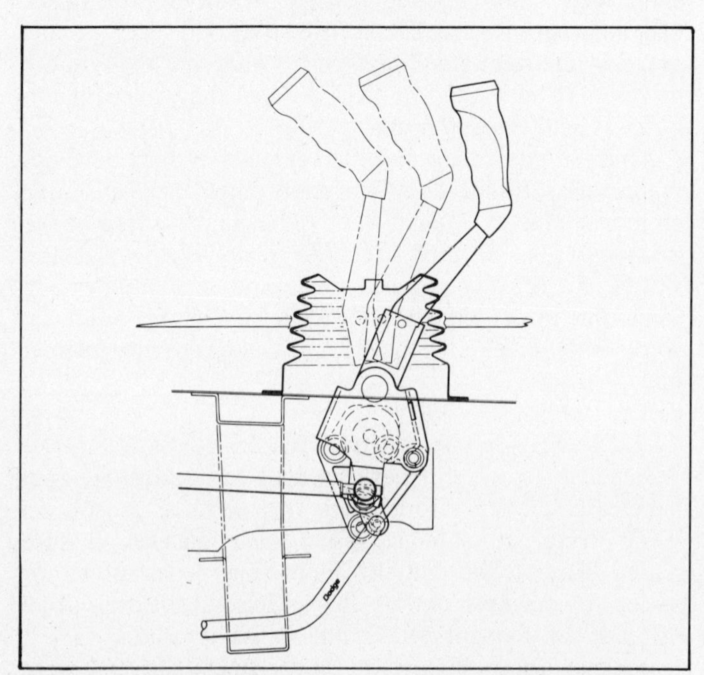

Chrysler designers went to extremes to dream up console shifters with character. This gun-grip shifter that Dodge offered in 1970 made the T-handle type look old hat. Drawing courtesy of Chrysler Corp.

I discussed the new four-speed manual transmissions from GM, Ford and Chrysler in Chapter 3. These were very popular on supercars until the late '60s when automatics gradually began to take over. But the four-speeds were upgraded during these years. Developments like a broader selection of gear ratios, improved floor shifters, stronger internal parts and lightweight aluminum cases improved them considerably. By the late '60s, clutches and manual four-speeds were quite durable in brisk street driving. The companies even warrantied them—though I'm sure they were a headache to dealer-service people.

More often than not, a salesman would try to sell you an automatic.

AUTOMATIC DEVELOPMENTS

As mentioned in Chapter 3, Chrysler engineers were the first to tailor a three-speed automatic transmission to take the punishment of all-out drag racing. It was an integral part of their '62—'63 Super-Stock program. They modified the A-727 Torqueflite with things like higher oil pressures for firmer shifts, and more clutch capacity with additional discs and upgraded friction material. The heavy-duty A-727 also had stronger splines, gears and shafts in the gearsets, heavy-duty bands and oversize sprags for reverse locking.

The remarkable results they achieved in racing with the Torqueflite encouraged other companies to beef up their automatics for their street supercars—though on a milder scale. By the late '60s, you could order a heavy-duty automatic with almost any street engine, including ones with mechanical valve lifters, capable of over 400 HP at 6500 rpm. Most notable were the Turbo Hydra-matic 400, first introduced in GM supercars in '65, and the Ford C-6, used behind '66-and-later Ford big-block V8s.

When you think about it, it's hard to believe the compa-

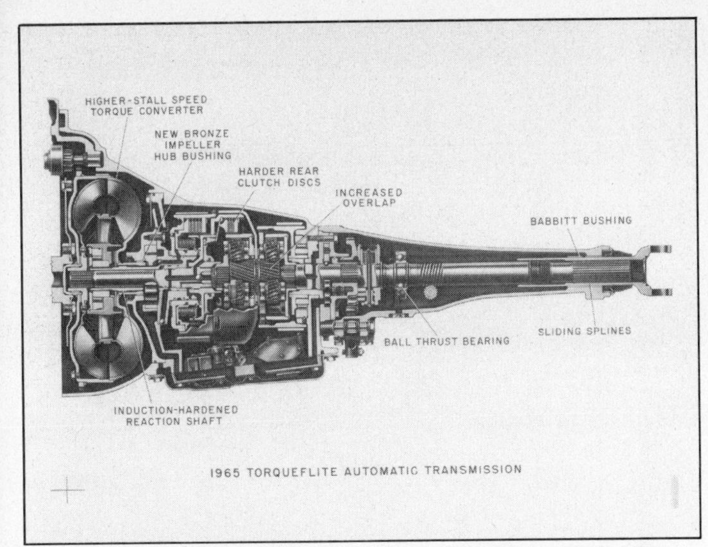

Chrysler was the first to offer a heavy-duty automatic suitable for drag racing with big-block engines. Originally offered in 1962, the A-727 Torqueflite was further upgraded for the new Hemi engine in 1965. Photo courtesy of Chrysler Corp.

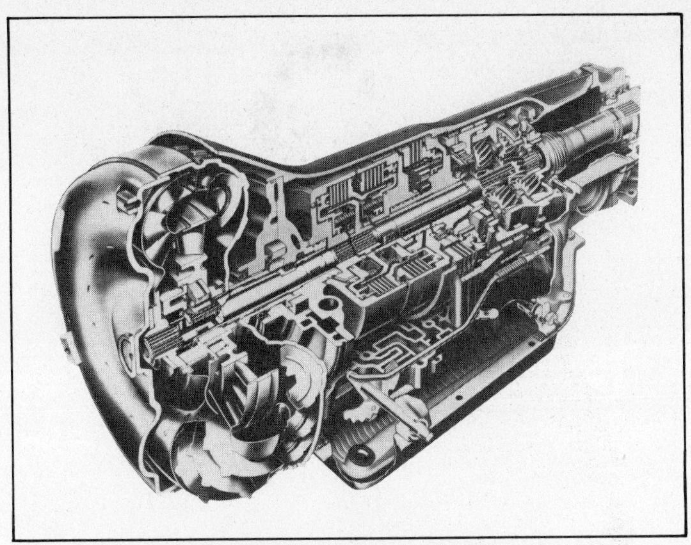

GM's first torque-converter three-speed automatic was a winner—the Turbo Hydra-matic 400. It was introduced in GM supercars in 1965. Capable of handling 500 ft-lb of torque as it came from the factory, the Turbo 400 remains the strongest passenger-car automatic GM ever made. Photo courtesy of General Motors.

nies could possibly warranty a combination like this. Admittedly, the transmissions were usually calibrated to upshift under 6000 rpm in **Drive** range or **D2** (intermediate) range. Many owners never knew you could hold in low in the **D1** position on all but some Turbo Hydra-matic 400 transmissions. Many different calibrations were used. I recall the solid-lifter '69 427 Chevelle had a Turbo 400 that would upshift at 6200 rpm in **Drive**—and I've seen more than one that would go to 6700 in **D1**. That was hard on engine and transmission alike.

The companies tried to protect themselves somewhat by calibrating automatic upshifts in **Drive** at a sane rpm. Probably most buyers stuck with this. But you *could* abuse an automatic by manually shifting it just like a four-speed. Many did. It's obvious there must have been a lot of profit in supercars for the companies to be willing to release such caged dynamite to the public. In fact, the warranty record of automatic-transmission supercars was remarkably good. The cushioning effect of the hydrodynamic torque converter, and the inate ability of the automatic's planetary gearset to make wide-open-throttle shifts without gear clash absorbed a multitude of sins!

Another vital automatic-transmission development in the late '60s was the *high-stall* torque converter. The stall speed of a converter is the engine rpm attainable at wide-open throttle when the output shaft is *stalled,* or prevented from rotating—like a *torque-up* start with the brakes locked. A *tight* converter in a standard car might have a stall speed of 1400—1600 rpm. But by designing the converter to have *looser* coupling, the stall speed can be raised to well over 2000 rpm. This helps off-the-line acceleration by allowing higher engine rpm where higher torque is developed. It's almost like having an extra low-ratio gear.

One unpleasant byproduct of a high-stall converter is additional slip under cruising conditions. This hurt gas mileage and created damaging heat. But with gas at 35¢ a gallon, no one worried.

By 1970, most high-performance automatics were using some form of high-stall converter. Stall speeds were in the 2000—2800-rpm range. Technology existed to go a lot further with this idea. But at higher rpm, slip becomes so bad that *tip-in* throttle response is sluggish. More-recent advances in torque-converter technology allow reasonably high stall speeds with little slippage at cruising speeds. But in the late '60s, such converters weren't available. Cruising slip was so bad on the Chrysler Street Hemi, for instance, that the MoPar boys adapted an external oil cooler from one of their trailer-towing packages to cool the transmission fluid!

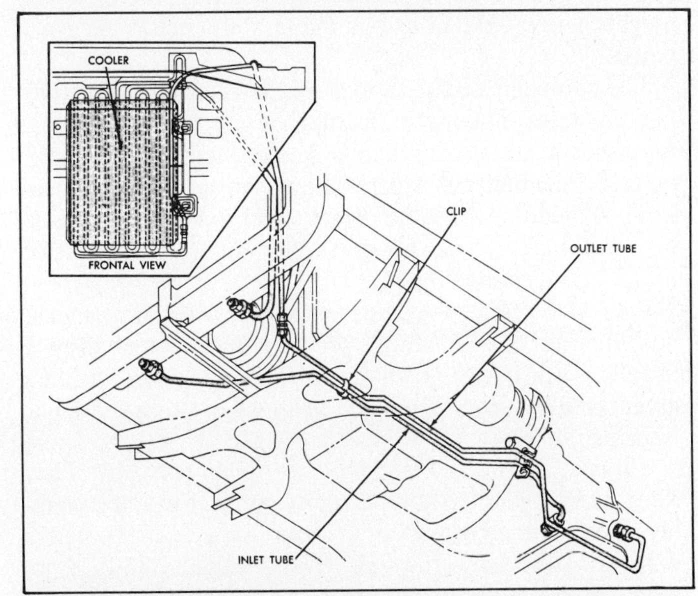

Chrysler engineers thought a separate transmission oil cooler was necessary on Street Hemis, because of the extra fluid slip with its high-stall torque converter. Durability was a real problem on supercars. Drawing courtesy of Chrysler Corp.

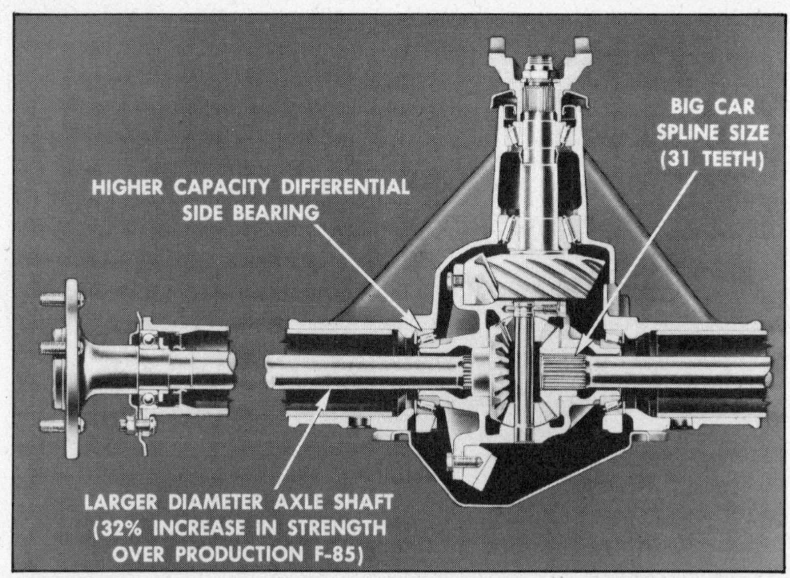

Typical of most Detroit supercars was a beefed-up drive line. These are the measures taken to ensure rear-axle durability in Olds 4-4-2s. Photo courtesy of Oldsmobile.

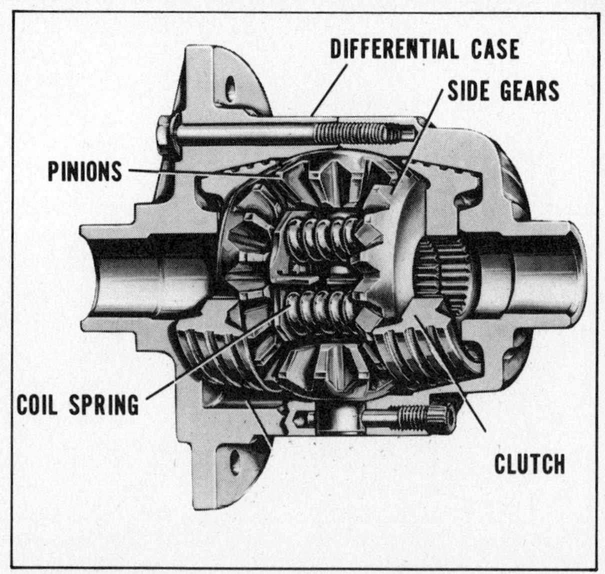

Many muscle cars came with limited-slip differentials to help control tire spin on hard starts. This is the Chrysler Sure-Grip type, with spring-loaded cone clutches. Photo courtesy of Chrysler Corp.

DRIVE-LINE BEEF

One good feature of an automatic transmission is the cushioning effect of the torque converter. Shock loads on the drive shaft, U-joints and rear axle on hard takeoffs and shifts are greatly reduced. So these components needed little attention in an automatic-transmission car, even behind a 400-HP engine.

On the other hand, manual transmissions were murder on drive lines. The problems multiplied as engines got bigger and stronger in the late '60s. Clutches were also getting larger, with more clamping force. And improved tire traction gave more holding power at the other end, compounding the problem. Something had to give. More often than not it was a U-joint, differential-pinion gear, or maybe a rear-axle housing. These were notorious weak points.

The companies used various methods to fight the drive-line breakage. The use of ductile-, or nodular-iron castings for critical pieces like U-joint yokes and rear-axle housings helped. This material was much-less brittle than plain cast iron, so it could withstand higher shock loads before breaking. Automakers use the stuff today for front-suspension uprights and connecting rods.

Rear-axle gears were sometimes strengthened by special heat-treating, or by the use of exotic nickel-steel alloy. It usually wasn't feasible to try to fit larger gears within a given axle housing. They had to make existing gears and housings stronger.

Chrysler solved the size problem in another way. They went shopping for a complete replacement rear-axle assembly. They ended up with a Dana *truck* axle: It had a 9-3/4-in. ring gear! The 9-3/4 Dana weighed 35-lb more than Chrysler's biggest automotive axle and cost a bundle. But you could slam the clutch and bang shifts all you wanted without breaking anything. It was standard equipment behind a manual transmission in Street Hemis and 440 Six-Packs. The Dana was optional fare in other Chrysler supercars.

To put the strength of these axles in perspective, the basic 9-in. Ford and 9-3/4-in. Dana rear axles are used in virtually all Funny Cars and Top-Fuel Dragsters running today!

SUSPENSION SCIENCE

By the mid-'60s, practically all supercars had some form of heavy-duty suspension package to match handling to the increased speed and acceleration capabilities. This usually consisted of higher-rate springs, stiffer-valved shocks, and a larger-diameter front anti-roll bar. Nothing particularly sophisticated here—though highway handling and cornering were usually improved measurably until you hit your first stretch of rough road. The mostly front-heavy V8 cars understeered considerably, but were predictable and stable.

As mentioned in the previous chapter, Oldsmobile pioneered with a *rear* anti-roll bar on the 4-4-2, greatly reducing the understeering tendency. The 4-4-2 was the best handling—read *most neutral*—of the supercars. Some of the other companies later went to the rear bar—or offered an over-the-counter kit. The cars almost invariably benefited from the change.

Attention was also given to controlling wheel hop and rear-axle windup. Chrysler used a shorter front section on their rear leaf springs, as everyone did eventually. They supplemented the asymmetrical springs with a rear-axle-pinion snubber that bottomed against the body underside. It was very effective. Some 427-CID Fairlanes used a separate rear-suspension torque link. Chevrolet and Ford used staggered rear shocks with the leaf springs on many late-'60s and early-'70s cars.

Actually there wasn't enough tire traction available to require a whole lot of attention to rear-suspension-torque control. But with today's tires and yesterday's engines, this would be a *real* problem.

BRAKES

Highway-patrol officers have known it for years: If you drive fast, you need more-effective brakes. It just makes sense. In fact, it was common in the '60s to fit police-pursuit brake linings on high-performance cars as standard equipment. These linings were made of a harder, semi-metallic material that could take more heat without fading. They required more pedal effort when cold. But most supercars had power brakes, so this was no problem.

The problem of pedal effort did, however, do in the infamous sintered-iron linings of the late '50s and early '60s. These were so hard that even a vacuum booster could hardly provide a decent stop on a cold morning. They were great for racing when they were kept hot, but horrible for the street. The factories eventually dropped them.

Actually the industry-wide trend to front disc brakes in the late '60s did more for overall brake performance than any of the previous gimmicks. *All* cars benefitted. Supercars still got slightly harder, fade-resistant friction pads. But it was no longer a critical problem.

Disc brakes on front wheels became popular during the late '60s. These '67 Mustang discs helped minimize fade during high-speed braking. Corvette even went to four-wheel disc brakes in '65. Photo courtesy of Ford Motor Co.

TIRES

Tires are important components on any high-performance car. Simply put, they are the only contact between the car and the road. Their grip determines what the car can do in terms of cornering, braking and straight-line acceleration. All the power in the world can't do anything without tires to transfer the force to the road.

In fact, tires were a major *limiting* factor in the performance of early supercars. Rubber smoke was everywhere. The best street tire you could get in the early '60s was the Atlas *Bucron*—made of a butyl-rubber compound that seemed to be a littler stickier than the standard rubber. In the mid-'60s, the best-performing street tire was the Goodyear Blue Streak. But, of course, passenger-car tires were very narrow in those days, putting only 4 or 5 in. of tread width on the road. Not much bite was available under the best conditions.

A major breakthrough came in 1967 with the introduction of the 70-profile Firestone *Wide Oval* tire. Wide Ovals offered a lower, wider appearance. Section height-to-width—aspect ratio—was 0.70, with 1- to 2-in.-wider treads for a given load capacity than previous 80-profile tires. This extra tread width had a beneficial effect on grip. The wider section-profile tire mounted on a 1-in.-wider rim gave more cornering power and much-better handling.

Within one year, practically all Detroit's high-performance cars were using 70-series wide-tread tires. They didn't have bite comparable with today's sophisticated radial-ply performance tires. But they were a huge improvement over the old Bucrons of the early '60s. And when the tire companies introduced *bias/belted* construction in the late '60s, with a stablizing belt under the tread, supercar trac-

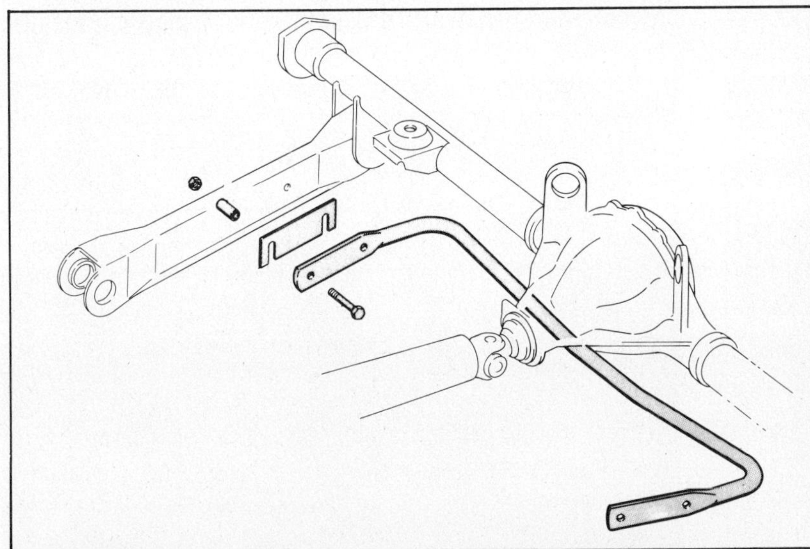

Olds had such good results with the anti-roll bar on the rear suspension of the 4-4-2 that some other companies offered the rear bars for dealer installation, or in do-it-yourself bolt-on kits. Drawing courtesy of American Motors.

Most of the '60s muscle cars had well-developed heavy-duty suspensions that gave decent cornering and handling—on smooth surfaces. But Detroit has made a lot of progress in this area in the last decade.

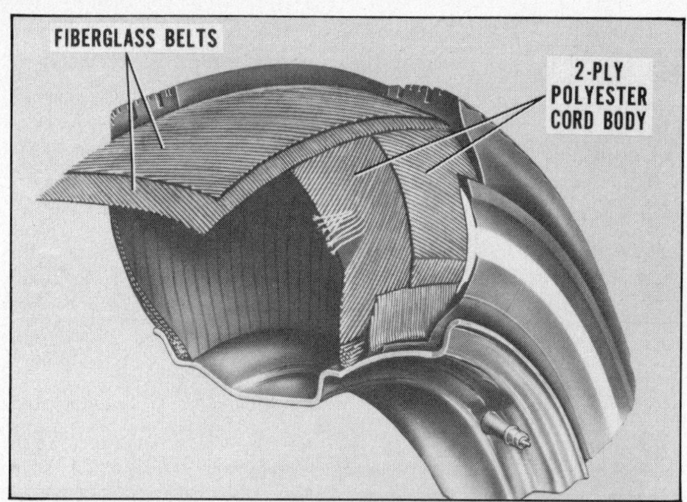

Bias/belted tire construction was introduced in the late '60s. It was an intermediate step between conventional bias-ply and full-radial construction. Traction and tread wear were improved.

Before Detroit got the hang of it, supercars were often big-engine monsters with Plain-Jane interiors. This is a view of the Spartan '66 Dodge Coronet Street Hemi. Photo by Roger Fehland.

The low-profile or wide-tread tire was a vital breakthrough in the late '60s—first with 0.70 height/width aspect ratio, then with 0.60. The added tread area and lower profile helped traction as well as handling. This Goodyear Polyglas GT was one of the first. Photo courtesy of Hurst Performance, Inc.

tion got a little better yet.

The proof is in the e.t.: The hottest supercars could turn drag-strip times in the 13s on wide-tread tires, each with less than 7 in. of tread on the track!

CONSIDERING THE DRIVER

I have to sympathize with the designers who tried to individualize the supercar driving compartment. They were working within the constraints of the standard mass-produced instrument-panel and control layout. Plus there was never much budget to cover any tooling changes required for maybe only 20,000 units a year. No way could the supercars enjoy the luxury of an exclusive instrument panel, control layout and seating designed specifically for the performance enthusiast.

In fact, it's amazing how much individuality the designers *were* able to achieve, considering the limitations. They concentrated their efforts in three major areas:

(1) Instrument cluster: It proved simple to slip in a special *rally* instrument cluster in place of the standard cluster, integrated in the standard-panel layout. Typically it includ-

1962 Pontiac instrument panel had no room for a tachometer. So, the location of dealer-installed tachs varied. Photos by Wally Chandler and Ron Sessions.

Corvair Spyder Turbo had a sanitary instrument panel and good gage layout. This one even sported gages for cylinder-head temperature and manifold pressure. Photo by Ron Sessions.

Pontiac introduced the outside hood-mounted tachometer in the late '60s on Firebirds and GTOs. This American Motors design is combined with the fiberglass hood scoop.

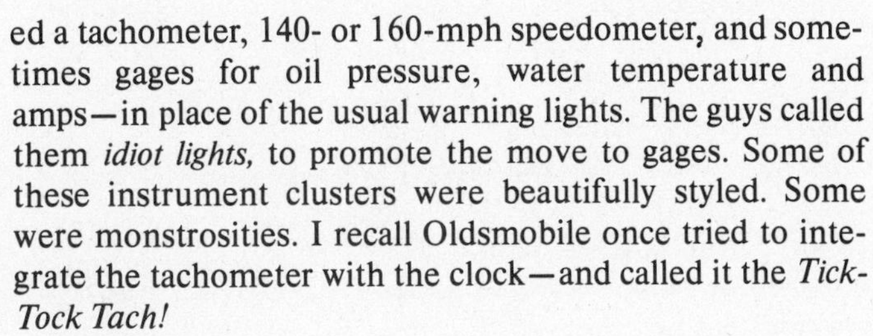

ed a tachometer, 140- or 160-mph speedometer, and sometimes gages for oil pressure, water temperature and amps—in place of the usual warning lights. The guys called them *idiot lights,* to promote the move to gages. Some of these instrument clusters were beautifully styled. Some were monstrosities. I recall Oldsmobile once tried to integrate the tachometer with the clock—and called it the *Tick-Tock Tach!*

(2) Steering wheel: It was a perfect component for customizing. The special wheel could bolt in place of the standard one. And needless to say, a variety of styles showed up—good and bad. Remember the Road Runner beep-beep horn and cartoon inset? Generally, the special wheel would have three or four open spokes, with a hub flashing some type of medallion. Many had some form of extra padding on the rim. This could be imitation leather with fake lacing or maybe black sponge rubber. There were many styles, some of them very impressive.

(3) Console: Because many supercars were sold with front bucket seats as standard equipment, a center console was a natural. The image-boosting console simply bolted onto the transmission tunnel. A console proved a good vehicle for several items. For one, it could be the base for a fancy shifter for either a manual four-speed or an automatic transmission. There was a great variety of shifters beautifully styled into the console: T-handles, trigger-grips, and so on. Also, consoles were often used to mount small tachometers, where tooling budgets didn't justify a special instrument cluster. Testers for magazines would always criticize console-mounted tachs because they were small and out of the line of sight. But it was either this or no tach at all.

In fact these little problems with instruments and driving controls only emphasize the total concept that made the mass-produced supercar possible. That is, it was essentially a custom high-performance car, made by bolting a few special parts on a standard bread-and-butter passenger car. This is why you could buy more performance-per-dollar in the U.S. than anywhere else in the world.

Optional "racing" steering wheels and custom instrument clusters were popular on high-performance cars of the '60s and '70s. They usually cost extra, but dressed up the interior a lot. This is a '70-vintage Dodge Challenger. Photo courtesy of Chrysler Corp.

The console of this '64 Chrysler 300-K included a manifold-vacuum gage. In this Chrysler, this gage was called a *Performance Indicator.* In the mid-'70s, the same-type gage was called an *Economizer!* Oh well. Photo courtesy of Chrysler Corp.

Richard Petty won the Daytona 500 in 1964 with the new Chrysler Hemi engine—the first race it ever ran. It developed at least 50-HP more than the big-block Fords and Chevys.

6

Race Cars on the Street

NASCAR made 'em do it

CHRYSLER STARTS A NEW GAME

Race fans watching the thundering super stocks of the early '60s—409 Chevys, 406 Fords, 413 MoPars—had no idea those engines would be like kiddies' toys compared to the monster being prepared by Chrysler for the 1964 NASCAR season. I'm referring, of course, to the fabled 426-CID Chrysler Hemi. This engine, built off the B-block, shared only its name with the '50s-vintage Hemis. I'll call it the *second-generation Hemi.* From the day it debuted at the Daytona 500 in February, '64, it dominated the NASCAR tracks, big-engine classes at the drag strips and many classes at the boat races. The big Hemi dominated practically every type of motor racing that required a *stock-block* engine. This was the premier factory race engine of all time.

You must understand the dynamics of early-'60s NASCAR racing to see how the second-generation Hemi could have happened at all. The key to the whole thing was the fact that early-'60s NASCAR Grand National rules did not specify a minimum production run to qualify a factory-built engine as *stock.* As long as it was built by a U.S. car company, whether in a toolroom or on an assembly line, it was a *stock* engine.

Chrysler pushed this concept to the logical limit with the Hemi. Their idea was to use the wildest, best-breathing heads on the 426-CID production cylinder block without changing camshaft location. This allowed using some existing tooling without overly compromising breathing. Chrysler also tooled up other major components, where it was obvious the existing pieces couldn't stand the gaff of racing. Special parts exclusive to the Hemi included bigger forged connecting rods with larger bolts, a high-output oil pump, forged counterweighted crankshaft, forged pistons, cross-bolted mains and a deep oil sump. Even the block casting was beefed up. As it turned out, only a few parts were retained from the production 426-wedge B-block engine.

Sure, it cost a lot of money. But NASCAR race-track victories were considered important in selling cars in those days. Chrysler planned to build only 20—30 Hemi engines a year—only enough to supply a select group of professional racers. They were to be built more or less by hand in factory-prototype shops. No massive tooling or million-dollar assembly lines. Just a quiet little toolroom job. All the noise would be on the race tracks.

Most car fans are well familiar with the unique cylinder-head layout of the Hemi. The combustion chambers were true hemispheres, with the sparkplug near the center of the dome and the valves inclined across the engine. Big valves were operated by rocker arms facing in opposite directions on two parallel shafts. The pushrods came up almost between the cylinder bores.

The layout had several important performance benefits.

Above: Plymouth built a number of lightweight Super Stocks in 1964 with the new 426-CID Hemi race engine. They were highly successful on drag strips for many years.

Ford SOHC 427 with cam cover removed. In a word, awesome. Photo courtesy of Ford Motor Co.

When Ford wasn't permitted to use the new single-overhead-cam 427 on NASCAR tracks, they shoehorned it into some Comets and Fairlanes to run in the NHRA A/FX class. This is Paul Rossi's intimidating '65 Comet.

The central sparkplug and hemi-type combustion chamber gave good combustion with high compression. Inclining the valves made room for very large diameters: 2.25-in. intake and 1.94-in. exhaust. There was no shrouding around the valves to impair flow. And, most important, the inclined valves gave large, smooth ports with very little curvature. Breathing was almost ideal with the layout. The only real disadvantages were that the inclined spread valves made the engine wider and heavier. And, of course, the valve-train arrangement was expensive to produce in volume.

Anyway, the new Chrysler Hemi engine was apparently right for racing. From the time it hit the NASCAR tracks, it developed 50—75-HP more than the wedge-chamber Fords and Chevys, inch for inch. This advantage carried over into drag racing as well. The hemi-type combustion chamber and valve layout was once again the darling of the racing world.

NASCAR BLOWS THE WHISTLE

Within days after the Chrysler Hemi engine hit the NASCAR tracks, Ford top brass gave their engineers the green light to develop an even more-powerful and exotic *stock-block* engine. This was the famous single-overhead-cam (SOHC) 427 Ford, first seen at the NHRA Winternationals in February, 1965.

Ford followed much the same pattern as Chrysler did with the Hemi. They took the existing 427-CID NASCAR block, with its side-oiling system and cross-bolted main caps, and developed entirely new hemi-type cylinder heads. But Ford went a step further: They operated the inclined valves through rocker arms from single camshafts running along the *top* of each head. The overhead cams were driven by a complicated roller-chain arrangement at the front of the block. This cam-drive setup was mounted on a large aluminum plate bolted to the front of the block. A lightweight magnesium cover housed the cam drive. A dummy camshaft in the original location drove the distributor and oil pump. The SOHC 427 Ford was a clever bolt-up arrangement. Any 427 NASCAR block could be converted with the necessary parts.

Actually, the new Ford heads didn't breathe a whole lot better than the Chrysler Hemi heads. But Ford engineers were depending on the reduced valve-train inertia to allow the engine to turn 7500—8000 rpm—and thus beat Chrysler with higher revs. It was also hoped the absence of lifters and pushrods would allow more-radical cam-lobe contours with greater lift rates for more power.

Well, none of it really had a chance to work out. When NASCAR officials got wind of the Ford overhead-cam development, they moved quickly and decisively. They felt they had to stop this trend toward hand-built factory race engines. The rulemakers had regretted allowing Chrysler to run the 426 Hemi in 1964. They felt this circuit should be confined to engines and cars somewhat near showroom configurations—especially as GM had recently opted out of the racing picture. If Chevrolet and Pontiac couldn't develop competitive factory race engines, it would be strictly a Ford/Chrysler show.

Thus, late in the '64 season, the announcement came down: Starting with the '65 season, any engine for the Grand National circuit would have to be installed in a minimum of 500 cars and sold to the public through new-car dealers.

Needless to say, this bombshell triggered some head-scratching around the Ford and Chrysler diggings. The thought of hand-building 500 of these exotic engines and installing them in cars staggered everyone. Ford officials admitted that early examples of the overhead-cam 427 cost them $14,000 apiece. They opted to forget the NASCAR scene. Instead, they threw the SOHC engine in a handful of Mustangs, Comets and Fairlanes for drag racing in the FX classes. In fact, they won the A/FX and B/FX trophies at the '65 NHRA Winternationals, first time on the strip. But Ford built less than 30 of the new engines that year. The SOHC was never factory installed in a production street machine.

A few Ford SOHC 427s found their way into roadsters and dragsters in the late '60s. Though power was equal to the Chrysler Hemi, the *cammer* wasn't very successful because of lack of parts and factory development. Photo by Warren Tanzola.

Street Hemi engine was offered in a variety of bodies in the '66—'71 period, with some 9000 cars being built. This is one of the first ones, a '66 Dodge Coronet Hemi. Photo by Roger Fehland.

Mighty Hemi filled the engine compartment of this '66 Dodge Coronet. Photo by Roger Fehland.

Chrysler's reaction to the NASCAR 500-unit ruling was initially one of outrage. It was especially frustrating after they had been allowed to race the Hemi for a season. Chrysler's racing chief, Ronnie Householder, did the logical thing. He withdrew all the Dodge and Plymouth factory-race cars for the full '65 NASCAR season. Ford factory cars won all the marbles by default with the 427 wedge engine.

But all this had a very positive effect on the supercar scene: The magnificent Dodge/Plymouth *Street Hemi* engine came out of it.

After the initial shock of the NASCAR 500-unit rule, Chrysler officials calmed down and began to look at the picture logically. It was obvious NASCAR wasn't about to cave in and allow hand-built race engines. Plenty of independent race teams were making decent showings with the Chrysler wedge engine. Ford had canceled the overhead-cam-engine program, and were reaping lots of publicity with the old 427 wedge because of Chrysler's pullout. What's more, Chevrolet was beginning to sneak back into the picture with a brand new Mark IV big-block with staggered valves and unique *semi-hemi* combustion chambers. Its image at stake, Chrysler could't afford to turn its back on NASCAR racing.

So midway through the '65 season the decision was made to put the Hemi back on the race track by *putting it on the street.* The decision was made to invest as much as was needed to re-engineer and detune the engine for everyday street driving. But the Chrysler people also envisioned another promising use for the muscular Hemi combination—as a high-output marine engine for heavy, high-performance boats. It was a natural for this application, with its excellent top-end breathing and rugged bottom end.

Chrysler officials actually got enthusiastic about the future of the engine. It was decided to tailor the machining and assembly facilities in Chrysler's marine-engine division to produce up to 3000 Hemis a year, for street, marine and race-track use. The Hemi was converted from a full-race engine to a high-performance production engine.

With the production facilities in place, Chrysler didn't have to try to squeeze by on the minimum requirement of 500 Hemi-equipped cars. The *Street Hemi* package was announced for a broad variety of 1966 Dodge and Plymouth models using the intermediate B-body—including the Dodge Charger, Coronet, Plymouth Belvedere and Satellite. In the next six years, through the 1971 model run, nearly 9000 Street Hemi-equipped MoPars were sold to the public. The Street Hemi could be had at package prices in the $900—$1100 range, depending on the mandatory equipment that had to be ordered with the engine. It was certainly one of the better horsepower-per-dollar bargains of the supercar era.

The Street Hemi wasn't the only race-bred car/engine combination put on the street to satisfy the NASCAR

External view of the Street Hemi shows neat streamlined exhaust manifolds and dual 4-barrel carburetion. Engine was available with either automatic or four-speed manual transmission. Photo courtesy of Chrysler Corp.

Because the Hemi cylinder heads didn't have exhaust-heat crossover passages, exhaust heat for the intake manifold was brought up through steel tubes from the right exhaust manifold. It worked. Photo courtesy of Chrysler Corp.

500-unit rule. The Ford Boss 429 Mustang of the '69—'70 period was produced specifically to make the 429-CID *Shotgun* or *Blue Crescent* engine eligible for Grand National racing. NASCAR minimum-production rules were a little more flexible than NHRA rules. Whereas NHRA required that the exact engine/body combination had to be available to the public, NASCAR rules specified only that the engine had to be produced for the street. This loophole allowed Ford to satisfy the minimum-engine-build rule by shoehorning Boss 429s into sporty Mustangs, while racing the potent powerplant in intermediate NASCAR Torinos and Montegos.

The Plymouth Superbird and Dodge Daytona *streamliners* were also examples of responses to the 500-unit rule. To make these special bodies legal for the speedways, Chrysler sold hundreds to the public.

NASCAR racing had a tremendous influence on high-performance development in the '60s. Companies were willing to spend millions in special tooling to produce those 500-or-more units necessary to make a specific combination *stock*.

Let's take a close look at these cars.

INSIDE THE STREET HEMI

It wasn't any simple job for Tom Hoover and his crew of Chrysler performance engineers to civilize the 426-CID Hemi race engine. To make it driveable on the street, five specific areas were addressed:

(1) Compression ratio: To enable the engine to use premium pump gas, lower piston domes were used. Compression ratio dropped from 12.5:1 to 10.5:1. Pistons were still forged.

(2) Cam timing and valve train: Low-speed torque was enhanced by using a milder-grind camshaft with 276° duration and 0.460-in. lift. Mechanical lifters were retained. Lower-rate valve springs gave about 280 lb of force at full lift. The dual-spring assembly was retained.

(3) Exhaust manifolds: Open, tubular-steel exhaust manifolds were discarded due to noise. New *tuned* cast-iron exhaust manifolds were designed with separate primary passages. They fit in the production B-body chassis.

(4) Heated intake manifold: To maintain cold-weather driveability along with efficient induction, a new aluminum intake manifold was used. On top were two 650-cfm Carter AFB carburetors. A two-plane layout and very large passages, or runners, were used. An exhaust-heat chamber was cast into the manifold to warm the mixture during cold operation. This chamber couldn't be fed by the usual crossover passage in the manifold because the Hemi heads were designed for racing: They had no provisions for such passages. Instead, Chrysler brought exhaust heat from the right manifold through two steel inlet and return tubes. It was kind of a Mickey Mouse setup, but it worked. You *could* operate a Hemi in cold weather.

(5) Progressive throttle linkage: An elaborate method of staging the throttle operation of the two 4-barrel carbs assured decent response and fuel economy under cruising conditions. Up to 60 or 70 mph, cruising was done on the front barrels of the *rear* carb. When these throttles were 40% open, the front barrels of the front carb started to open. Meanwhile the rear secondary air valves of both carbs were controlled strictly by airflow—they gradually opened as flow through the primary barrels increased. When all eight barrels opened—in the 3000—4000-rpm range—power came thundering on. The tremendous howl through the single low-restriction air cleaner could be heard three blocks away.

To the delight of enthusiasts, many of the exotic race parts were retained for the Street Hemi. The engineers knew most Street Hemis would be raced one way or another, and they wanted to provide reasonable reliability under these conditions. After all, Chrysler had to warranty the engine. Goodies common to street and race versions included a forged-steel crank, cross-bolted main-bearing

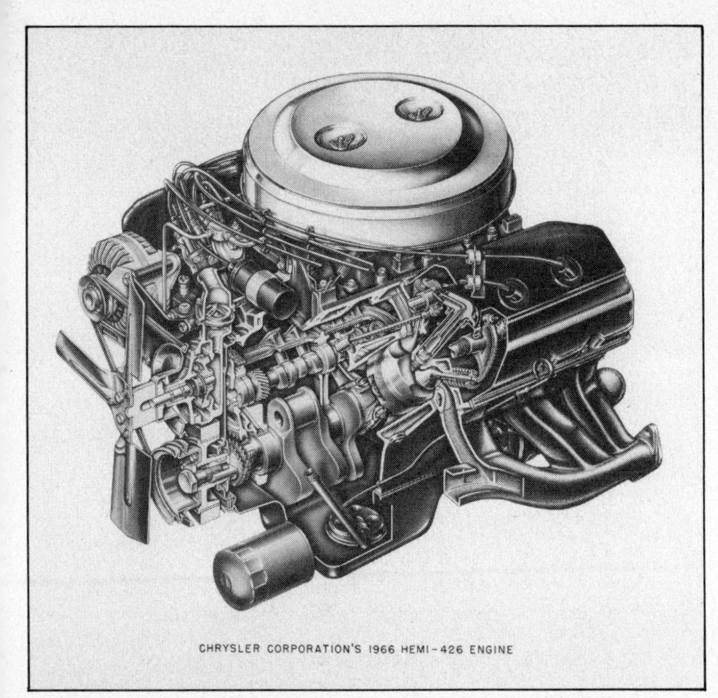

Chrysler engineers did a clever job of detuning the wild Hemi race engine for street use in 1966. This cutaway shows the unique inclined-valve layout with two rocker shafts. Big heads made it a very heavy engine, though not as heavy as the original '51—'58 Hemi. Photo courtesy of Chrysler Corp.

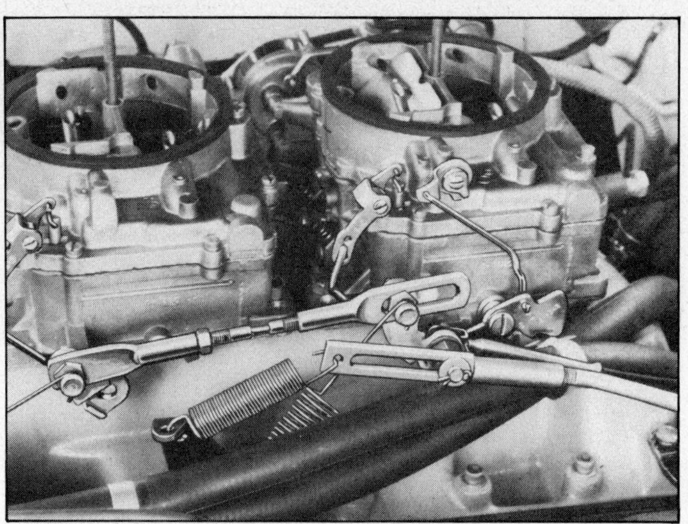

Street Hemi had a unique *staged* throttle linkage. While cruising, only the front barrels of the *rear* carb were used. As wide-open throttle approached, the front primaries, and secondaries of both carbs opened. Photo courtesy of Chrysler Corp.

In the mid- to late '60s, this little fender emblem was the only external identification on the Dodge or Plymouth Street Hemi. The engine did its own talking!

caps, beefier rod forgings with 7/16-in. bolts, chrome-alloy-steel valves, roller timing chain, high-output oil pump, deep-sump oil pan, and a special water pump.

Much attention was also paid to the drive line and chassis. Changes and additions to parts that were part of the Hemi-engine package contributed substantially to its price. Special parts included an 11-in. heavy-duty clutch, beefed-up Chrysler A-833 four-speed, Dana 9-3/4-in. rear axle and Sure-Grip differential. The optional automatic was a heavy-duty A-727 Torqueflite with a four-pinion planetary gearset. Also included were heavy-duty four-ply police tires, 11-in. heavy-duty police brakes with hard linings, high-rate springs and stiff-valved shocks.

With all this exotic hardware under the skin, the only external identification for the Street Hemi package was a little fender medallion. It was a throwback to the days of the early-'60s super stocks with monster engines lurking under seemingly tame bodywork!

HEMI ON THE STREET

No other American supercar ever generated as many myths and misinformation as the fabled MoPar Street Hemis. They were held in awe by the average street racer—who could hardly hope to scrape up the necessary $4500 or so to buy one. Most never had a chance to even ride in one.

One of these myths concerns the fabulous rev range of the engine. Forget it. With the *street* valve springs giving only 280 lb of force at full lift, and high reciprocating weight of the huge valves and long rocker arms, those early Hemis would peak at or below 6000 rpm—even with solid lifters. The Hemi Torqueflite was conservatively calibrated to upshift at 5500 rpm in **Drive** range.

Important improvements were made to the engine in 1968. A more-aggressive camshaft was designed: 284° duration, 0.480-in. lift, and improved lobe profiles to reduce valve-train shock loads at high rpm. Single valve springs replaced the duals, but gave about the same 280-lb force at full lift. Inner damper coils were used to reduced valve-spring *surge.* Bore sealing was improved with moly-filled top piston rings. Long-life valve-stem seals reduced oil loss. Finally, a windage tray under the crankshaft reduced whipping of the oil. The tray was said to add 15 HP at 6000 rpm. Add another 10—15 HP for the new camshaft and those '68—'69 Street Hemis were 30-HP stronger than the *Stage-I* '66—'67 engines. Additionally, the '68—'69 Hemis could rev to 6300—6400 rpm.

Another important change came in 1970: *hydraulic lifters.* The '70—'71 hydraulic cam had the same timing and lift as the earlier solid-lifter cam. But performance was improved because valve lash was eliminated. Chrysler engineers found that many Hemi owners were neglecting periodic valve-clearance, or lash, adjustments with the solid-lifter cams, so power gradually fell off as valve timing and lift changed. Wear and tear on the valve train was also a problem with solid lifters. Hydraulic lifters solved all of this.

Street Hemi engines were available in Plymouth Road Runners of the late '60s. A real sneaky combination, as these cars normally had relatively mild 383-CID wedge engines.

Street Hemis were popular in the new E-body ponycars in 1970, because of lighter weight and shorter wheelbase. This is a '70 Hemi 'Cuda.

Model	0—30 mph	0—60 mph	1/4-mile e.t. @ mph	Est. Net HP
'66 Plymouth Satellite	2.9 sec	7.1 sec	14.5 @ 95	330
'67 Dodge Charger	2.7	6.4	14.2 @ 96	360
'68 Plymouth GTX	3.0	6.3	14.0 @ 101	390
'69 Dodge Charger	N.A.	5.7	13.9 @ 104	420

Engineers tried to maintain the rpm range by using high-rate valve springs with 320 lb at full lift. These allowed 6200-rpm upshifts—but the increased friction between cam lobes and lifters caused rapid cam and lifter wear, and gradual loss of power. This problem always plagued owners of late Hemis.

Street Hemi changes are illustrated dramatically in the above table by CAR LIFE road tests of '66—'69 models.

You can see gradual increases in net power from minor engine changes made through the years. But it's interesting to note that 0—30-mph and 0—60-mph times for the '68 car were very close to the '67, even though quarter-mile times were much better. This is due to longer cam duration and overlap. Low-end torque was not as good on the '68-and-later Hemis, though they were stronger from 4000 rpm up.

These figures, incidentally, are directly comparable because all test cars, except for the '69 Dodge Charger, were equipped with a Torqueflite automatic and 3.23:1 final drive. The '69 Charger used a manual four-speed transmission and 3.54:1 gears.

CAR LIFE testers got their best Street Hemi times with the '69 Dodge Charger, even though it was harder to get consistent launches, or takeoffs, with the stick-shift car. Once under way, the 3.54:1 gears helped. The testers got a best e.t. of 13.68 seconds at 104.82 mph with the four-speed—but a dozen previous runs varied from 13.70 to 14.12 seconds!

Some interesting experiments were also run with a '68 Plymouth GTX Street Hemi. To test the effects of final-drive-ratio changes on performance, they switched the standard 3.23:1 gears for optional 4.56s. They also used 9-in.-wide slicks to get more-consistent takeoffs. The engine was completely stock and the street exhaust system hooked up. You can see how much the Hemi was helped by the low-ratio ring and pinion—which let the engine rev higher on initial takeoff and in 4th gear:

Standard 3.23 rear	13.94-sec e.t. @ 103.21 mph
Optional 4.56 rear	13.43-sec e.t. @ 104.86 mph

Private owners who ordered their Hemis with the optional 4.30:1 and 4.56:1 gears were also more successful in street racing. Although it had gobs of low-end torque, the engine definitely wasn't comfortable at low and medium speeds. The big disadvantage of the low-ratio gears, of course, was fuel economy: How about 5—7 mpg?

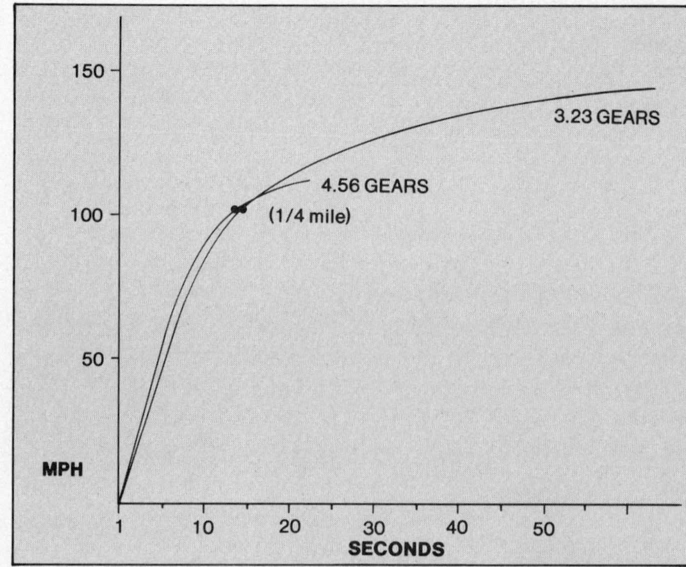

Street Hemi acceleration was especially sensitive to axle gearing.

Boss 429 street engine was a real shoehorn job in the '69 Mustang. In fact, the front-spring towers had to be moved outboard to get it in—a very expensive conversion. Asking price for the option didn't begin to cover Ford's costs.

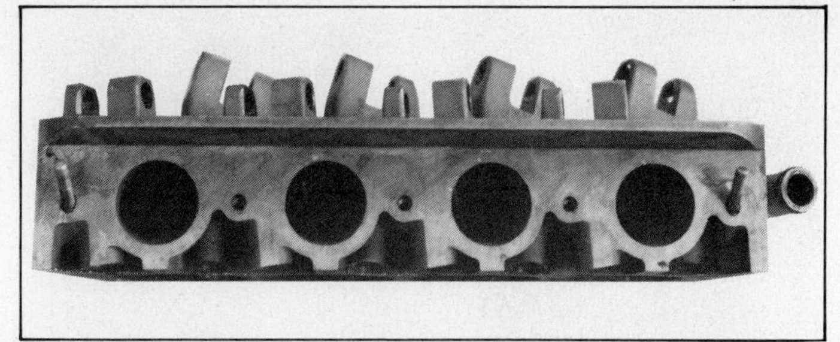

Boss 429 Ford engine had huge round intake ports with 4.3 sq in. of area. They were probably the biggest ports ever put on an American production engine—and definitely too big for the street!

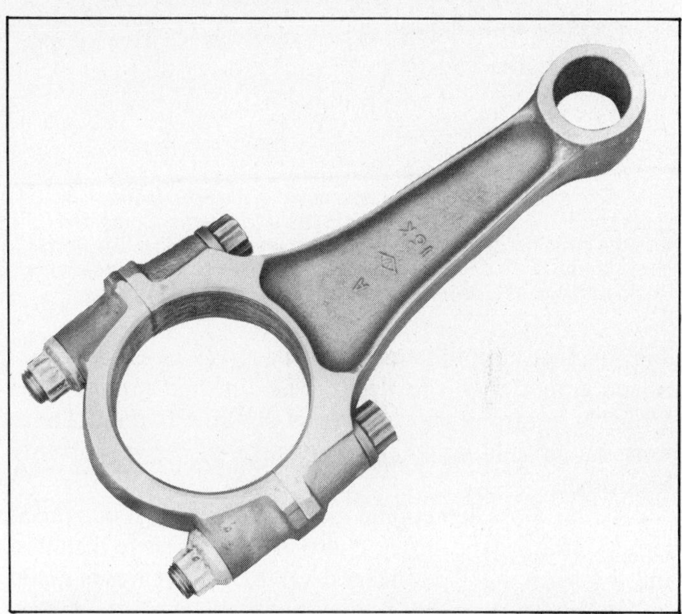

NASCAR rods for the Boss 429 Ford race engine were huge forgings with high-strength bolts. They were used on later Boss street engines.

A well-tuned Street Hemi with 3.23:1 final drive could approach a top speed of 150 mph. This required about 6000 rpm with stock tires. There was ample horsepower available to do it—if a tire didn't disintegrate first. Aerodynamic design was relatively non-existent by today's standards, so if it went 150 mph, it did it on brute power alone. Fantastic engine.

INSIDE THE BOSS 429 MUSTANG

The Boss 429 Mustang was *not* Ford's answer to the Street Hemi. Ford never wanted to build the Boss 429 Mustang—no more than Chrysler wanted a Street Hemi. Both cars were tremendously expensive to build. And they were expensive and troublesome to warranty—spare parts were an impossibility. Boss 429 Mustangs weren't practical to sell to the public—not to mention their balky street performance and terrible gas mileage.

But as I mentioned earlier, NASCAR made 'em do it. The infamous 500-unit rule of 1965 was the push behind these wild, semi-race street machines. If you wanted to run an engine *or* car body in Grand National competition, you had to produce at least 500 for sale to the public. Ford turned out 1360 Boss 429 Mustangs in the '69—'70 period to qualify the 429-CID *Shotgun* engine for NASCAR competition.

In case you're wondering, no, you didn't have to qualify a specific engine *with* a specific body. Mustangs weren't even allowed to compete in the Grand National class.

In some ways, this 429-CID NASCAR engine was more of a challenge to civilize than the Chrysler Hemi. For one thing, the intake and exhaust ports were much larger. You could practically stick your hand into those huge round intake ports. The 2.40-in. intake valves looked as big as saucers. Ford engineers liked the biggest possible ports in their race engines. These ports worked well above 6000 rpm, but on the street, velocity was so slow in the ports that the air/fuel mixture hardly knew which way to go! Fuel dropped out of the mixture, causing poor low-speed performance.

Ford's answer for the street Boss 429 was to choke down the induction system. For carburetion, a single 735-cfm Holley 4-barrel was mounted on a two-plane aluminum manifold with medium-size passages—but no exhaust heat. Intake-valve diameter was reduced to 2.28 in. A hydraulic-lifter cam provided a duration of 282° and valve lift of 0.510 in. Valve springs gave 310-lb load at full lift, allowing about 6200 rpm with hydraulic lifters. A lower 10.5:1 compression ratio was used with the hemi-type combustion chambers to help reduce detonation.

It wasn't a good combination for a street engine. But all agreed that Ford engineers did a good job of tailoring this all-out 7000-rpm speedway engine so it was reasonably tractable for driving around town.

Like the Chrysler Hemi, many bottom-end features were retained to give decent reliability up to 6500 rpm for street racing. These included: cross-drilled, forged-steel crankshaft, four-bolt main-bearing caps, forged pistons, dual oil galleries, and a high-volume oil pump and pan. Street Boss 429 aluminum cylinder heads were O-ringed to

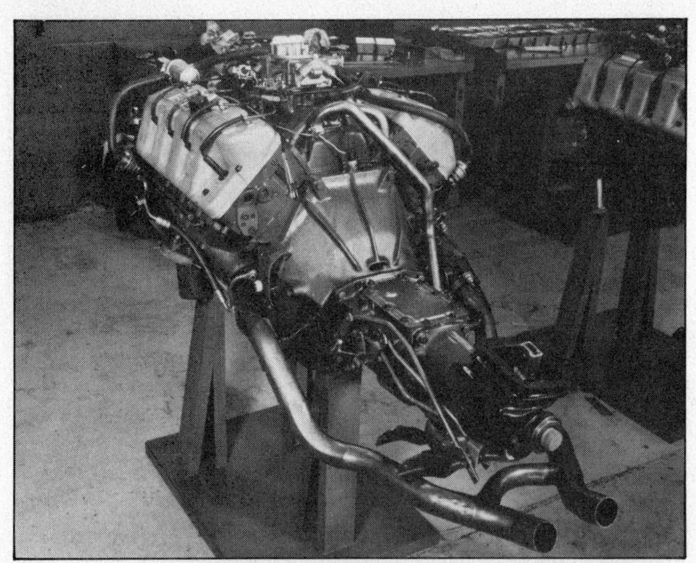
Boss 429 engines were hand-assembled in the Ford toolroom at great cost, with custom-fabricated exhaust systems. Combined with the chassis/body mods required to install the big mill in Ford's ponycar, the Boss 429 Mustang was the most-costly supercar ever. Photo by Thomas E. Stuck.

eliminate gasket problems. Race Boss 429 blocks were O-ringed rather than the heads. The finished product was probably as strong and reliable as the Street Hemi. There were no chronic weaknesses to cause excessive warranty headaches.

As if the Boss 429 engine itself wasn't enough of a problem to tame for the street, it also proved a bear to install in the '69 Mustang. The inclined-valve cylinder heads made the engine so wide that it wouldn't fit between the spring towers.

Ford could have installed the Boss in a Torino or Galaxie with a minimum of hassle. But they wanted it in the Mustang for image. This is the only way they felt the whole nightmare might pay off in youth sales. It meant engine installation could not be done on factory assembly lines.

What Ford did was to use their Kar Kraft special-operations facility in Dearborn to re-engineer the Boss Mustang. Spring towers were revised and moved outboard 2 in.—thereby widening the front track. Because the front-suspension control arms were relocated outboard, a new center steering link was used to maintain correct steering geometry. Heavy-duty tie-rod ends were used and the engine was installed on special mounts. Other goodies included heavy-duty suspension, heavy-duty four-speed Ford *top-loader* and heavy-duty 9-in. rear axle.

Kar Kraft controlled the production of the Boss 429 Mustang. Partially assembled Mustangs were shipped from the Dearborn Assembly Plant to a small plant in Brighton, Michigan, a few miles north of Detroit. Kar Kraft personnel completed these Mustangs, turning them into Boss 429s on their own assembly line.

Boss 429 Mustangs were a tremendous undertaking for Ford. The extra $1208 they asked for the package represented only a small fraction of what it cost to build those cars. They are to be commended for staying with the project long enough to produce more than 1300 of the monsters. NASCAR would have let them get by with only 500!

BOSS ON THE STREET

It would be nice if I could say the new Boss 429 went out and beat Chrysler's nasty old Street Hemi hands down. It didn't work out quite that way. As I said, those *Shotgun* ports were just too big for the street. Even with all the detuning and choking down by Ford engineers, they could not quite get the act together. Big ports made the low end and mid-range weak. And just about the time the big ports started to work on the top end, the restrictive carburetion and camming took over. The street Boss 429 didn't perform really well anywhere in the speed range.

Even so, the street Boss 429 was one of the quicker supercars of the period. CAR LIFE got these acceleration figures for a '69 four-speed with 3.91:1 final drive:

Ford's Boss 429 Mustang of 1969 was another street-engine combination forced by NASCAR rules. Ford lost thousands of dollars on each one produced. But they were fabled street machines. Photo courtesy of Ford Motor Co.

Boss 429 engine in street trim. Heated air intake improved cold-start driveability. Seal on air cleaner was for intake air. Photo courtesy of Ford Motor Co.

According to Ford public-relations people, the fastback roofline of '63-1/2 Ford was supposed to help aerodynamics on the NASCAR superspeedways. Photo courtesy of Ford Motor Co.

0—30 mph	3.2 sec
0—60 mph	7.1 sec
1/4-mile e.t.	14.09 @ 102.85 mph
Estimated net HP	370

In 1970, Ford engineers tried to upgrade the performance of the Boss 429 by going to a more-radical, 300° cam and solid lifters. To keep the bottom end together at higher revs, big NASCAR rods with 1/2-in. bolts were used. The combination did allow the engine to wind to 6500 rpm with the existing 310-lb valve springs. And there might have been another 10 HP above 5000 rpm with the longer-duration cam. But 1/4-mile acceleration performance wasn't helped a whole lot. Drag-strip e.t.s still hovered around 14 seconds, with trap speeds around 102 mph.

It was an interesting lesson in standing-start acceleration curves. With the street Bosses' huge ports and the soft low-end and mid-range torque, the car spent too much time getting up to 70 or 80 mph where flow in the ports was decent. At that point, there wasn't enough left of the quarter-mile to turn good numbers. All else being equal, the Street Hemi could get up to 60 mph 1- to 1.5-seconds quicker. And to top it, the street Boss 429 didn't have the carburetion and manifolding to haul on the top end. The single 735-cfm Holley and two-plane manifold were poorly matched to the size of the ports and valves underneath. Flow velocity was fine through the carb and manifold, but slowed at the big ports, hurting air/fuel mixture.

Interesting car.

AND EVEN STREAMLINERS

In the '50s and early '60s, the major emphasis in the development of NASCAR race equipment at the factory level was nothing more than brute horsepower. Everyone agreed that horsepower was what won races. Chrysler's success with the 426 Hemi and Ford's remarkable comeback in '66—'67 with the 427 dual-quad *tunnel-port* were conclusive evidence.

But there was another trend afoot in the mid-'60s that was changing this picture. This was the increasing popularity of high-banked super speedways on the NASCAR Southern circuit, where cars could hit speeds up to 200 mph. Tracks like Daytona, Talladega, Atlanta and Charlotte put new emphasis on high-speed performance. And, of course, this meant *aerodynamics.* Suddenly, the race mechanics began looking at the shape of the body—not only the power potential of the engine—when selecting and building a car.

I remember seeing publicity photos of the new '63-1/2 *Super Torque* Fords that compared the formal roofline of earlier models to the new semi-fastback Galaxie 500 XL/Marauder roofline. The side-view comparison showed reduced turbulence for the new models. Actually, this was a styling move that the Ford PR people jumped on to get more ink in the enthusiast magazines. Whether or not the semi-fastback roofline helped on the NASCAR tracks, I don't know.

In my opinion, Chrysler engineers were the first to give the race teams some special low-drag aerodynamics to work with. The Dodge Charger had been redesigned for 1968, with sleek *fuselage* styling and a *tunnel-back* layout at the rear window. Wheel-opening lips were flush. The quarter panel and roof C-pillars swept back, but the window notched down steeply between the rear roof pillars. Despite its racy looks, this roof-pillar layout was bad from an aerodynamic standpoint.

There was a lot of turbulence in the hollow behind the window. Chrysler's fix was to throw a gently sloped window across the gap between the C-pillars, and extend it back farther on the rear deck. The change could be made on the original Charger body line with relatively simple tooling. Additionally, the grille was moved forward, flush with the front end. This further helped the coefficient-

Pretty, they weren't. Fast, you bet! Specially streamlined Plymouth Superbirds were successful on the NASCAR speedways in 1970. Here is Dan Gurney testing one on the now-defunct Ontario Speedway road course.

Aerodynamics of the Superbird were suitable for another track—Bonneville. With the right gearing, these otherwise-stock "Birds" would go more than 200 mph. Photo by Tom Monroe.

1970 Daytonas and Superbirds were so good that Richard Petty was lured back to Chrysler after his defection to Ford the previous year. This is Pete Hamilton at Daytona. Photo courtesy of Daytona International Speedway.

of-drag (C_d) factor. The overall result was a big reduction in aerodynamic drag when combined with the rounded body sides and flush wheel-opening lips.

Chrysler's Charger 500s were released to NASCAR teams early in the '68 season. The *500* designation represented the number of cars Dodge built to make the model eligible for Grand National racing. The cars were said to be 2—3-mph faster than the original Chargers on the super speedways.

This was only the beginning. Encouraged by the track record of the 500 model in 1968, Chrysler engineers went much further in 1969. They set out to clean up the Charger body. Using scale models in the 7 X 10-ft wind tunnel at Wichita University, they designed a pointed nose for the 500 body that reduced the C_d by a full 23%. Admittedly, it wasn't very practical for the street. The new nose extended 18-in. farther forward and eliminated the front bumper. Try to get away with that today! Also, the headlights were concealed under vacuum-operated doors for a further reduction of C_d.

Just when the engineers figured they had the wind-cheating nose finalized, a strange thing was discovered. The aerodynamics of the tapered nose unloaded the rear tires. The fix was a unique rear wing to give downforce. The wing was supported at the ends by vertical struts 2-ft above the rear deck, Flash Gordon style! Craziest-looking thing you ever saw. It was an expensive body modification to produce.

But the modifications did the job at high speeds. The new Dodge *Daytona* appeared at least 6—8-mph faster than the '68 Charger 500s—with the same horsepower!

Bobby Isaac, driving Harry Hyde's Daytona, was the fastest man on the NASCAR speedways in 1970. Hyde had the knack for setting up top-end engines. And Isaac was usually the fastest qualifier on the banked tracks. His one-lap record of 201.10 mph on the Talladega oval still stands as this is written. A few months after this record, they took the car to Bonneville and set a bunch of national B/Production straightaway records. The car hit a top speed of 211.76 mph with all factory parts in the Hemi engine. Doesn't this say something about the supercar era?

And yes, these cars are most-valuable collectors' items today. Dodge made 800 Daytonas in 1969. Plymouth made about 1800 nearly identical cars in 1970, calling them *Superbirds*. Both came with the 440 Magnum wedge engine as standard equipment. Trim level was equivalent to Dodge

Dodge and Plymouth had to build at least 500 streamliners to make them eligible on NASCAR tracks. This is a street version of the '69 Dodge Daytona model with optional Hemi engine. Photo courtesy of Mopar Muscle Club.

Chrysler used the Hemi race engine in a number of special drag-racing models in the '65—'68 period. These had magnesium cross-ram dual-quad manifolds, and some had aluminum heads—good for 650 HP. This installation was in Ronnie Sox's '68 Barracuda. Photo courtesy of Holley Carburetor.

In '68, Ford introduced a truly fastback roofline on its all-new Ford Torino and Mercury Montego intermediates. But the car's nose had the C_d of a brick wall. Still, Ford pilots like Cale Yarborough did reasonably well that year. Photo courtesy of NASCAR.

R/T and Plymouth GTX models. The 426 Hemi was optional.

The inevitable question doesn't have a firm answer: How fast could one go with the right final-drive ratio and a stock Street Hemi engine? I once heard of an owner who did 6400 rpm with 3.23:1 final drive, running Goodyear track tires. That figures out to more than 160 mph. One thing is certain. There wasn't a *faster* American-supercar combination.

Another factory streamliner effort, though less well known than the radical Chrysler cars, were Ford's *Talladega* conversions of the Torino and Montego coupes in 1969. Ford's "better idea" here was similar to Chrysler's: Stick an extended sheet-metal nose on the front of a standard body to reduce front-end drag, and provide a minimum-opening grille to cool the engine on the street. Ford's nose wasn't tapered to a point like Chrysler's, so it wasn't as efficient aerodynamically. But it only extended 12-in. forward

In '69, Ford remedied the situation with the Talladega conversion. The nose sported a flush grille and much-reduced frontal area. LeeRoy Yarbrough won the '69 Daytona 500 with one. Photo courtesy of NASCAR.

Details of the 426 Hemi cross-ram manifold and huge Holley 4150 4-barrels. Photo courtesy of Holley Carburetor.

One of the 80 or so '68 Hemi Barracudas made by Hurst for drag racing in the SS/AA class. NHRA required at least 50 cars be manufactured to qualify for the Super-Stock classes. These were very successful. Photo by Ron Sessions.

and didn't need that hideous rear wing! And it was possible to use conventionally mounted headlights. The Talladega was a simpler and less-expensive conversion. With the new 429 Shotgun engine, Dave Pearson won the '69 Grand National points championship with his Mercury.

Incidentally, Ford made about 800 Torino and Montego Talladega conversions—most powered by the 428 Cobra Jet engine. They are extremely rare cars.

AND SOME NHRA RACE CARS

You may recall from Chapter 3 that the National Hot Rod Association stipulated a minimum production of 50 car/engine assemblies to qualify a model for the Super-Stock classes. Obviously this was a much-easier bogie to hit than NASCAR's 500-unit minimum.

You might think there were a lot more special hand-built race combinations produced for NHRA drag racing than for NASCAR track racing. Actually there weren't. Drag racing didn't attract nearly as much public attention in those days—nor did national drag-race victories seem to have as much influence on the youth market. The factories weren't willing to put as much effort and money into special packages for NHRA racing.

But there were a couple of interesting projects in the late '60s worth mentioning. When looking at these cars, keep in mind that there was no special effort to make them streetworthy. With only 50, 60 or 70 units to distribute, they could be channeled strictly to serious racers. A few found their way onto the streets, but this was the exception. In fact, many of the cars are still racing today—with only minor changes.

Dodge/Plymouth—For the 1968 NHRA season, Chrysler commissioned Hurst to build a series of A-body Plymouth Barracudas and Dodge Darts with 426 race-Hemi engines to run in the top four-speed and automatic S/S classes.

The engines had all the latest factory goodies, including twin Holley carbs on a magnesium cross-ram manifold. Bodies were stripped and gutted. Batteries were relocated in the rear to get better front-to-rear weight distribution. Cast-iron cylinder heads replaced the aluminum ones used on earlier race Hemis. The cars also had the big, 9-3/4-in. Dana rear axle with fabricated traction bars, super-duty springs and shocks, and a beefed drive line.

These were complete race cars just as they came out of the Detroit-area Hurst job shop. Their competition record has been brilliant ever since.

Chevrolet—Many supercar fans don't realize that Chevrolet built approximately 70 Camaros in 1969 with the all-aluminum Corvette big-block *ZL-1* competition engine. The aluminum 427 was 160-lb lighter than the cast-iron 427. These cars were not gutted or radically modified for racing. The basic idea was to get a better front/rear weight distribution with the aluminum engine, so the cars would be more competitive in one of the lower S/S classes. This was achieved.

The car had a curb weight of about 3300 lb with full street equipment. About 56% of that was on the front wheels—about equivalent to a Z/28 model with the small-block 302. Traction was good.

I had an opportunity to strip-test one of the cars in 1969, just as it came off the assembly line. This one had 4.10:1

There was no external identification for the ZL-1 427—but you could tell by the performance! This one has been restored to new condition. Photo by Bill Porterfield.

final drive, close-ratio Muncie four-speed "rock-crusher" and F70 street tires. All we did was unhook the headpipes from the exhaust manifolds. The best run was a 13.16-second e.t. at 110.21 mph, shifting at 6500 rpm. That suggests a net output of considerably more than 400 HP as tested. NHRA didn't accept the 430-HP advertised rating—and factored it at 470 HP. Chevy claimed more than 500 HP with tubular-steel headers. That Chevy ZL-1 427 was healthy!

Incidentally, very few ZL-1 Camaros found their way onto the drag strips. The aluminum block didn't work out well for 8000-rpm speeds. Most of the cars either ended up on the street, or the engines were pulled and sold to Can-Am, circle-track or boat racers. Maybe the owners felt they had to convert some of the hardware into cash: Chevy put a hefty $4965 price on the ZL-1 package! Including the car, that added up to a whopping $8134 with manual transmission or $8224 with an automatic!

American Motors—The AMC people became interested in drag racing their Javelin ponycar in the late '60s. In 1970, they produced a limited run of AMX two-seaters for SS/D competition using the 390-CID V8. The cars were sent to the Hurst job shop for gutting and fabricating.

In those days you had to use a *factory* intake manifold in the S/S classes. So, AMC put a factory part number on an Edelbrock competition intake manifold and Holley carb. The top competitor in the SS/D AMXs was Shirley Shahan, one of the early woman drag racers. She did well by AMC in the late '60s and early '70s.

Chevrolet installed all-aluminum ZL-1 427 engines in 70 Camaros in 1969 to run in the SS/C class. They weren't successful drag racers—but these cars are popular collector items today. Photo by Phil Chisholm.

And so it goes. In the American-supercar era, you were just as apt to see a factory race car on the street as on the race track.

Top Mustang performance option in '64-1/2—'66 was the 271-HP 289 "HiPo." Photo by Ron Sessions.

This is the car that started the *ponycar* stampede—the Mustang. Shown are Ed Giola's '66 GT 2+2 and '65 *K-model*—HP289—convertible. Photo by Ron Sessions.

7

Ponycar Revolution

Sports cars for the masses

FORD'S BETTER IDEA

When Ford Motor Company introduced the Mustang in 1964, it was a marketing coup. But car fans didn't realize that this low-priced, four-seat ponycar would also prove to be a breakthrough in performance and handling. The performance advantage was nothing more than a matter of weight. Here was a compact-size car that weighed 200—300-lb less than conventional intermediates of the period. And handling and maneuverability promised to be better because of the shorter 108-in. wheelbase. Handling was helped because of improved weight distribution gained by moving the engine and seats back a little.

Everything was a step in the right direction for performance. But Ford officials didn't really have hard-core performance in mind when they laid out the Mustang in the early '60s. This car was born of what seemed to be a brand-new market demand.

Lee Iacocca took over the Ford presidency in 1960, at the age of 36. One of his early ambitions was to bring out a new car to appeal to a broader segment than the lukewarm Falcon compacts and Fairlane intermediates of the period. He wanted a low-priced car, but something that could be dressed up with a big variety of options to satisfy a range of tastes. Iacocca wanted something on the sporty side, but nothing exotic. In fact, there was a lot of internal pressure to bring out a two-seat successor to the popular '55—'57 Thunderbird. But Iacocca insisted on four seats and decent trunk space.

You know what resulted. The Mustang was a compact car with a base price of $2368. You could make it into practically anything you wanted with the option list. Its long-hood, short-deck silhouette set a styling theme.

All agreed this new look was a vital factor in the success of the car. The car had character and image from Day One. It was attractive to performance fans and young lady office workers and retired businessmen all at once. It was a car that couldn't miss.

In fact, Iacocca's predictions proved to be more restricted than his dreams. He projected sales volume for the new car at 100,000 units a year. But from the day the showroom doors opened in April 1964, production was back-ordered up to four months. Sales in the remainder of calendar 1964 totaled an astonishing 680,000 units!

Mustang became the most-successful all-new model to come out of Detroit at the time.

The Fight Is Joined—Needless to say, within a few weeks after the Mustang hit the street, other Detroit companies launched crash programs to develop their own copycars. It was similar to the way the industry copied Pontiac's GTO. But it took longer because they had to tool up completely new cars.

Above: Original 1964-1/2—'66 Mustang started the trend to long-hood, short-deck styling. This rare '65 K-model is equipped with the 271-HP 289 V8, pony interior and red-stripe tires. Photo by Ron Sessions.

Most car fans and Mustang enthusiasts alike agree that the original '64-1/2—'66 Mustang had the cleanest lines. Photo by Ron Sessions.

"HiPo" 289 V8 nestled neatly into the engine compartment of this '65 convertible. Note extra tubular braces used to add front-end rigidity. Photo by Ron Sessions.

Today, a major redesign can be done in two years, with the help of computer analysis and computer-aided design. In those days, it took more like three years.

Chrysler was first to answer the Mustang challenge with their quickly cobbled Barracuda. They got wind of the Mustang project early and rushed the Plymouth Barracuda into production. Introduced only a few weeks after Mustang's debut, the '64-1/2 Barracuda was mainly a fastback roof transplanted on a plain-Jane Valiant body—and it looked it despite Chrysler styling director Elwood Engle's best efforts. The first *real* Barracuda, designed by Chrysler Imperial stylist Dave Cummings, came out in '67.

Chevrolet did remarkably well to get the new-from-the-ground-up Camaro into production by the summer of 1966. Pontiac's Firebird, using most of the same tooling, came on stream about six months later. Ford also introduced the Mercury Cougar about this time; essentially a dolled-up, long-wheelbase Mustang. Then in the fall of 1967, American Motors got their Javelin coupe going—plus a shortened 97-in.-wheelbase version, the AMX, touted as a genuine high-performance sports car.

None of the other ponycars enjoyed anywhere near the sales success of the Mustang. In fact, even Mustang sales dropped off substantially after 1966 as the other brands ate into what appeared to be a fairly fixed market segment. By the early '70s, government regulations, rising insurance rates and fuel-octane problems definitely reduced market demand for high-performance, sporty cars.

The most-popular ponycar model in the mid-'70s was Pontiac's Firebird *Trans-Am*—selling upwards of 100,000 units in '78 and '79. Pontiac capitalized on the Trans-Am competition image as others dropped out of the market.

The ruboff worked so well that Pontiac was and is willing to pay SCCA a royalty for each car produced to use the name. Pontiac officials had the good sense to doll up the looks with decals, spoilers and fender flares, making the car look different from anything else on the road. It caught the fancy of car fans across the country, becoming sort of a poor-man's Corvette. Performance and handling, though good, were secondary considerations.

You might say Ford, Pontiac and perhaps Chevrolet were the only companies to make huge profits on ponycars. The real legacy left by this type of car was that it set new standards of performance and handling during a period when conventional supercars were getting bigger, heavier and clumsier.

As mentioned earlier, this was accomplished by careful weight control—helped by unitized-body construction. Handling was also aided on most ponycars by moving the engine and seats back to get better weight distribution. The long-hood, short-deck styling theme was the indirect cause of this.

It worked! All the ponycars, except the early Barracuda, had similar silhouettes. And wheelbase lengths fell in the narrow range of 106—111 in. The cars were remarkably similar in overall concept.

Ford obviously hit on a great idea with the first Mustang. The other companies were not ashamed to copy it almost line for line.

Here's a brief rundown of how these individual ponycar models evolved . . . .

PARADE OF PONYCARS

Ford Mustang—The original '64-1/2 Mustang was available in a notchback coupe and convertible. A *2+2* fastback was introduced a few months later. All were small, lean-looking cars, with no excess fat. Their *curb weight*—full tank of gas and no passengers—was less than 3000 lb with the base six-cylinder engine.

There was definitely no room for a big-block-V8 option. Top performance option in those early Mustangs was Ford's 289-CID small-block V8, introduced the previous

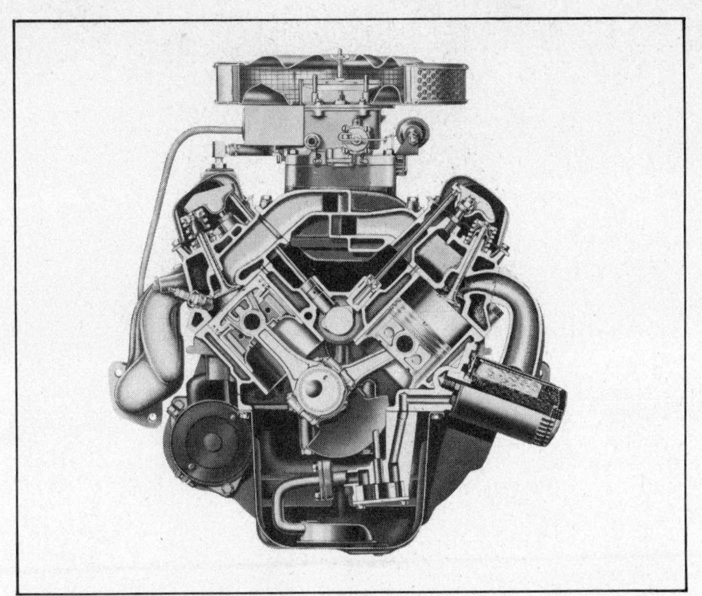

High-performance 289-CID V8 featured high-compression heads, a high-output cam, 4-barrel carb, streamlined exhaust headers, and so on. An exciting little engine. Photo courtesy of Ford Motor Co.

Mustangs of the late '60s with the optional 428 Cobra Jet engine were some of Ford's strongest supercars. Plenty of power, light weight and short wheelbase got the job done. Ram air was optional. Photo courtesy of Ford Motor Co.

Action at the 1977 NHRA World Finals pits a '69 Mustang 428 CJ against a '69 Mustang convertible. Photo by Tom Monroe.

year in the Fairlane intermediates. A special high-performance version of this engine, the High Performance 289, featured 11.5:1 compression, 480-cfm 4-barrel Autolite carburetor, open-element air cleaner, a long-duration, solid-lifter cam and high-rate valve springs. The *HiPo* 289 also featured a stronger crank, rods and pistons, thicker main-bearing caps and a special low-restriction dual exhaust system.

Ford rated the HiPo 289 at 271 HP at 6000 rpm—and it would definitely go to 7000 rpm or more without any distress in the valve train. It was an exciting little engine, with its unusual revving ability and the high-pitched roar from the open-element air cleaner. It remained a popular option in several lighter Ford models through 1967.

The HiPo Mustang was more than a high-performance engine. It had upgraded drive-line components such as a heavy-duty clutch, Ford's top-loader four-speed manual transmisson with close-ratio gears and a standard 3.50:1, 9-in. rear axle. A modified C-4 automatic transmission was also available. It was calibrated to shift at 6500 rpm at wide-open throttle. Special chassis components in the High-Performance package included: disc brakes, high-rate springs, stiff shock absorbers, fast-ratio steering gear and red-line Firestone tires. All these goodies—engine included—sold for $750 extra! The High Performance Mustang was an economical, flexible little sports/muscle car.

By late 1965, word leaked out that the coming GM pony-cars would have large-enough engine compartments to accommodate big-block V8s. This was the signal for Ford engineers to spread the front-spring towers and track width of the Mustang to accommodate their large engines.

Redesigned '67 models were about 2-in. wider and weighed about 100-lb more with similar equipment. According to plan, the wider front compartment allowed the 390-CID big-block V8 as an option in standard or *GT* form (335 HP). Then in 1968, the 427-CID *hydraulic* engine (390 HP) became available. Later still, the magnificent 428 Cobra Jet took over. This was the strongest street engine ever offered in the Mustang. And definitely one of the quickest showroom models of the supercar era.

Some of the most-exciting Mustangs were produced in '69—'70. These include the Boss 429 and the Boss 302. In my opinion, the '69—'70 Mustangs were the cleanest designs since the original '64-1/2—'66 version. These cars had crisp, taut lines.

Ford took it upon itself to redesign the Mustang for 1971. The resulting cars were almost a foot longer, nearly a half a foot wider and 300—400-lb heavier than their predecessors. And most car fans, even the most-dedicated Mustang lovers, will agree that the '71—'73 Mustangs were styling failures.

In 1971—'72, Ford made a limited number of Mustangs with high-performance 351-Cleveland engines. In '71, they were the Boss 351 and 351 CJ. In '72, the Mustang was available with a 351 HO and a high-performance four-

Major styling change in 1971 added considerable weight and size to the Mustang. Added to the loss of power from tightening emission controls, the increased weight killed performance. Photo by David Gooley.

Cobra Jet version of the 351 Cleveland engine was a good powerplant for the Mustang in the early '70s. But Ford got out of performance before the combination had a chance to be developed. This is the '71 Mustang Boss 351.

barrel 351 with flat-top pistons and 9:1 compression to run on low-lead fuel. The '71 330-HP Boss 351 featured 11.7:1 compression, solid lifters, four-bolt mains, dual-point distributor, windage tray, low-restriction exhaust manifolds and an optional fresh-air intake system.

In 1971, Ford also built a *Cobra Jet* version of the 429-CID 385-series V8. But these were never generally available. Performance development at Dearborn practically stopped after "Bunkie" Knudsen left in 1970. The 428 CJ Mustangs of the '68—'70 period remain the pinnacle of Mustang straight-line performance.

Plymouth Barracuda—The first Barracudas don't meet my strict definition of a ponycar: long-hood, short-deck styling with 2+2 seating. Yet, they were marketed as ponycars and considered by many to be ponycar contenders.

The Formula S model, introduced in 1965, was a serious performance effort. In addition to special identifying trim, the package included heavy-duty suspension, bigger drum brakes, wider 14 X 5-1/2-in. wheels, Goodyear Blue Streak tires, fast-ratio steering and an instrument-panel-mounted tachometer.

The Formula S engine was very special, too. You recall Chrysler brought out their modern small-block V8 (patterned after the Chevy small-block) in 1964. The first version displaced 273 cubic inches. It had small ports and an unusual single-plane intake manifold. The Formula S package included: Carter AFB, 10.5:1 compression, and a high-output cam. Its low-restriction exhaust manifolds fed into a single, large-diameter exhaust system with a "cowbell" tailpipe outlet. The combination was rated at 235 HP at 5200 rpm—up from 180 HP for the standard 2-barrel engine. It could rev to about 5600 rpm with standard valve springs.

At $258 more than the base Barracuda V8, the Formula S package was one of the best performance buys of the period. Also, this car was one of the best-handling ponycars, thanks mainly to the Goodyear Blue Streak tires. These Goodyears were way ahead of any other domestic tire at the time in terms of steering response and cornering power. Blue Streaks were also popular for police pursuit and sports-car road racing. Chrysler engineers were stick-

1971 Mustang Boss 351 was Ford's last real performance model until the 302 HO of '82. Photo by Ron Sessions.

1971 Mach I was the first and last Mustang available with the 429 CJ. Photo by Ron Sessions.

Ram-air ducting of '71 Boss 351 Mustang. Ducts carried outside air from NACA ducts in hood to airbox over carburetor. Photo by Tom Monroe.

When the Mustang was upsized for '71, it was done with an eye towards installing the big 429/460 V8. Photo by Ron Sessions.

In '65, Plymouth put a high-performance 235-HP version of the 273-CID V8 into the Barracuda, and made a *Formula S* model. It had outstanding acceleration and handling. Photo by Benyas Kaufman.

Plymouth spun the fastback Barracuda off their pilgrim Valiant—and it looked it. It was introduced within weeks of the original Mustang. Photo courtesy of Chrysler Corp.

ing their necks out to use them on a showroom car as they produced a noisy and harsh ride. But they worked.

In 1967, the Barracuda got its very own sheet metal and interior. Although it shared its underpinnings with the pilgrim Valiant/Dart, styling was crisp and contemporary. When the high-performance 340-CID V8 became available in '68, Barracudas suddenly became very hot street machines.

High-performance 273 V8 introduced in Barracudas, Valiants and Darts in '65. Higher compression, 4-barrel carb, high-lift cam and low-restriction exhaust bumped output to 235 HP. Photo courtesy of Chrysler Corp.

The 340 V8 is a story in itself. It was much more than a bored 273/318 with a few special parts bolted on. Chrysler engineers looked at every piece of this small-block engine to see what needed to be upgraded to handle 300 HP at 6500 rpm.

Barracuda underwent a major restyle in '67, adding a new coupe model. 340-CID version of the Chrysler small-block V8 was introduced in '68. With plenty of power on tap and decent weight distribution, these were potent road cars. Photo courtesy of Chrysler Corp.

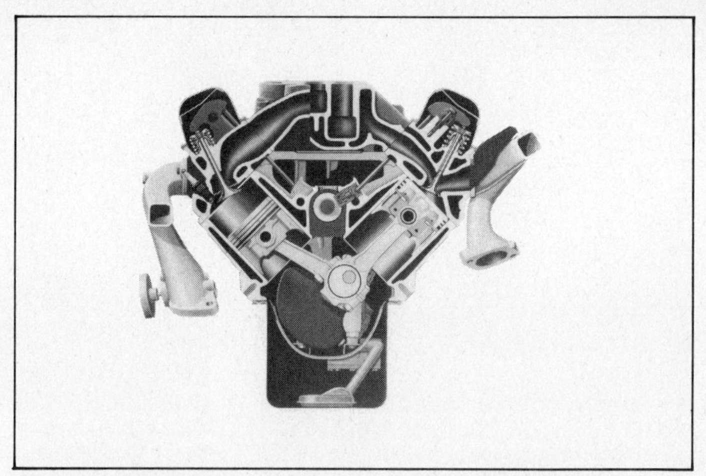

Sectional view of high-performance 340 V8 shows port layout, rugged bottom end and streamlined exhaust manifolds. Photo courtesy of Chrysler Corp.

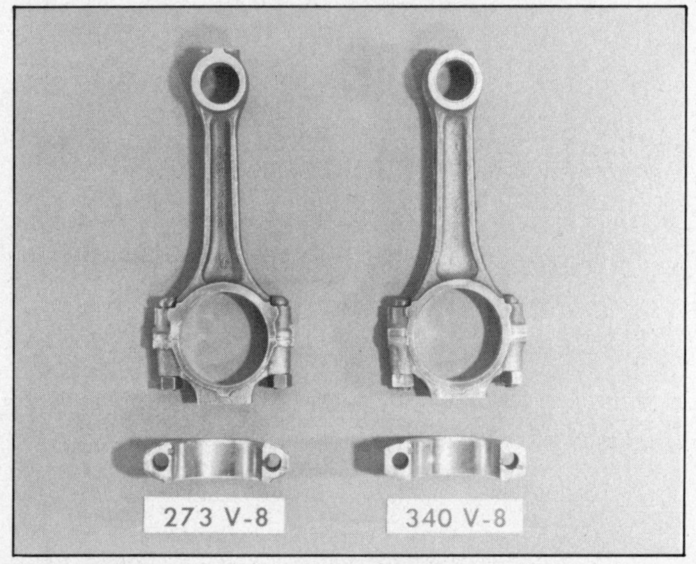

High-performance 340 V8 had high-strength connecting rods to allow 6500 rpm speeds. Standard 273/318 rod is at left.

Plymouth first offered the 383-CID big-block V8 in the 1968 A-body chassis. The '69 *383 S* shown had good acceleration, but the additional weight up front made the cars understeer badly. Photo courtesy of Chrysler Corp.

New cylinder heads had much-larger ports and valves. A new high-riser intake manifold mounted a 600-cfm Carter AVS 4-barrel. The new camshaft was supplemented with high-rate valve springs, heavy-duty pushrods and rockers. The bottom end was beefed up with a special-alloy forged-steel crankshaft, heavy-duty rods and oil pump, and a windage tray. External goodies included a low-restriction air cleaner, declutching fan and tuned, low-restriction exhaust manifolds.

Output of this engine, as installed in the car, was no disappointment. Chrysler rated the engine a conservative 275 HP at 5000 rpm to get the cars in more-favorable drag-racing classes. NHRA wasn't fooled: The cars were factored at 290 HP. According to my formula the 340 S engine, as installed in the A-body cars, had a net output of 290 HP. As such, it represents one of the few cases where a manufacturer's advertised gross-horsepower rating was actually *below* its real-world net-horsepower rating!

The 340 would wind comfortably to 6000—6200 rpm with hydraulic lifters. It had smooth, flexible response in city traffic, and gas mileage was 16—17 mpg at 70 mph on the highway! It was one of the sweeter street engines of the supercar era—but with plenty of top end.

Also, in 1967, Chrysler made the B-block 383-CID V8 available in the Barracuda. It was a shoehorn job, but at an advertised 330 HP, it promised to be a potent performer. As could be expected, the heavy block made the car very front-end heavy, upsetting its fine handling balance. And the big, thirsty engine didn't rev as freely as the small-block 340.

Dodge Challenger/Plymouth 'Cuda—By the late '60s, Chrysler was convinced the ponycar craze was here to stay. They decided to give Dodge a Barracuda sibling: It was dubbed the Challenger. But this time Chrysler's timing was off—way off. By the time their all-new E-body entry hit the showrooms in late 1969, increasing government regulations and rising insurance rates were dulling the youth/performance segment of the market.

But the new Challengers and 'Cudas were fully competi-

1970 Plymouth 'Cuda was too late to get in on the best part of the ponycar market. It was a good performer, but not a good seller. Photo by David Gooley.

Dodge/Plymouth E-body ponycars of 1970 offered a complete line of engines from the lowly slant six up to the mighty Hemi. This '70 Challenger had the "middle-of-the-road" 340 V8. Photo by Ron Sessions.

In '70, Dodge extended their popular R/T performance package to the new Challenger ponycars. Lighter than its Coronet and Charger counterparts, the Challenger had a performance edge. Photo by David Gooley.

Camaros of the late '60s became popular drag-strip material because of their light weight and short wheelbase. Photo by Ron Sessions.

tive in terms of performance. Street fans were uneasy with published shipping weights 200—300-lb more than similar Ford and GM ponycars. But apparently, the excellent dual exhaust systems introduced on these cars offset any weight handicap.

Some say the extra weight was a result of widening and beefing the chassis to accept the Chrysler big-block engines, including the mighty 426 Street Hemi. Could be! It's also true that the industry concept of the ponycar had taken on plushness and luxury in the late '60s. Even the 1971 Mustang was 9-in. longer and 500-lb heavier than the original 1964 version. At the end of the supercar era the ponycars were definitely getting too fat for optimum performance.

Chevrolet Camaro—Chevrolet's entry in the ponycar sweepstakes hit the showrooms in the fall of 1966. From the very beginning, Chevy engineers provided space and chassis beef for big-block V8s. This brought the Camaro in 200—300-lb heavier than the early Mustangs.

Model-year 1967 also marked the introduction of Chevy's 350-CID version of the venerable small-block V8. Also, like Mustang, those early Camaros were available with a couple of economy six-cylinder engines, small-block V8s displacing 302, 327 and 350 cubic inches, plus the 396-CID version of the Mark IV big-block.

The Camaro was in the showrooms only a few months when Chevrolet released the potent L78 version of the 396 to do battle in the horsepower race. The L78 featured big-port heads, 11:1 compression, big Holley carb, solid-lifter cam and a 6500-rpm bottom end. With the necessary driveline beef to back up the L78, Camaros so equipped weighed only about 3600 lb. Acceleration was impressive to say the least. And if you're wondering how the 375-HP L78 got around the GM corporate policy of not less than 10-lb per horsepower, the option was listed as dealer-installed. Yet GM brass looked the other way and allowed L78s to be dropped into Camaros on the assembly line!

Of course, the most-popular high-performance Camaros

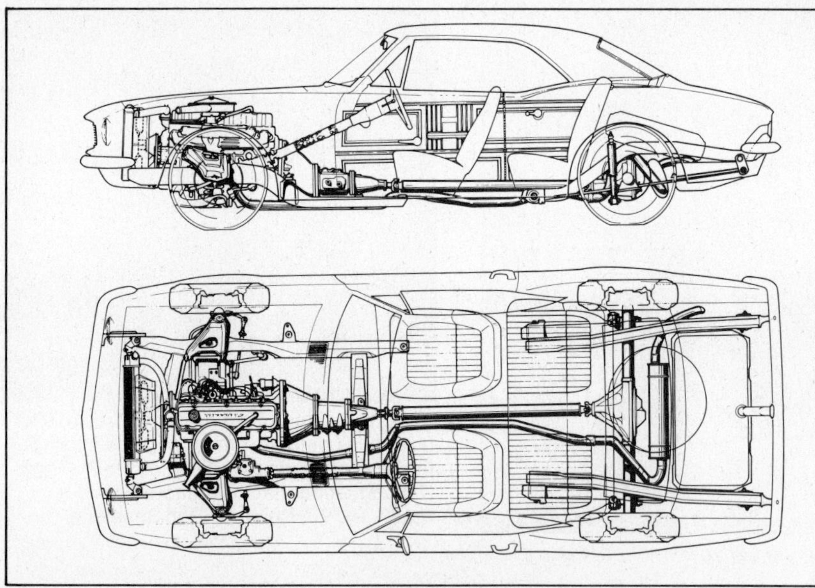

Chevrolet's Camaro ponycar, introduced in 1967, was a little bigger and more spacious than the original Mustang. But it stuck with the long-hood, short-deck silhouette that was proving so successful. Drawing courtesy of Chevrolet.

By 1968, the Camaro SS was available with the small-block 350 or the big-block 396. Photo by Ron Sessions.

When equipped with the Chevy small-block V8, Camaros had decent front/rear weight balance and good handling characteristics. Photo courtesy of Chevrolet.

Once the mono-leaf rear suspension was discarded in favor of multi-leafs, Camaros became popular on the drag strips. After all, they were Chevys!

of the supercar era were the *Z/28* models. I will discuss these later in the Trans-Am section. But for now, I'll just mention that they used the 302-CID high-performance engine—derived from using the 283 crank with 327 pistons—in the '67—'69 period. The 350-CID *LT-1* powered the Z/28 in '70—'72. These 350-CID Camaro Z/28s were quicker than the 302s, though not as much as you'd think. The extra cubes helped mainly to give more torque and peppier throttle response for normal street driving.

The Camaro was extensively redesigned for the 1970 model year, with big changes in sheet metal and trim, and some sub-frame and suspension improvements. These changes helped ride and handling, but weight increased another 150—350 lb—not good for acceleration. The L78 Camaros of the late '60s remained the quickest of the breed. Also, Chevrolet made about 150 Camaros with 427-CID engines.

Pontiac Firebird—The Firebird appeared about six months after the Camaro. Though the cars shared essentially the same chassis and body shell, they differed in sheet-metal, trim, engines and some drive-line details. The Firebird weighed 100—150-lb more than the Camaro, model for model, probably because of fancier trim and interiors and additional sound deadening.

The extra development time allowed Pontiac to work out some of the bugs in the basic Camaro design. For example, where Chevrolet struggled with axle-hop problems in mono-leaf, rear-suspension Camaros, the Firebird came with multi-leaf rear springs from the start. And where Chevrolet ended up placing gages in the center console to leave room for a dash-mounted tach, the Firebird broke new ground with a hood-mounted tach! Using the same basic instrument panel, the Firebird had gages where they should be—in front of the driver.

Camaro underwent a major restyle for 1970. This is the '72 SS 350. Photo courtesy of Chevrolet.

Chevrolet stylists put a lot of effort into the Camaro interior, trying to appeal to a broad market. This is the '70—'77 *cockpit* instrument-panel layout. Photo courtesy of Chevrolet.

Like most ponycars, Firebirds offered an impressive lineup of engine options. In the first three years of production, the unique Overhead-Camshaft Six was available. Pontiac engineers put a belt-driven, overhead-cam cylinder head on the 230-CID Chevrolet-L6 block and touted it as a poor-man's Jaguar. The combination was available with high compression, 4-barrel carburetor and dual exhaust. It could wind to 6500 rpm with no distress. The OHC Six didn't have the horsepower or torque to deliver supercar acceleration, but the engine made some wild sounds. Firebirds so equipped were definitely a ball to drive.

If you wanted more power, you could opt for a 326-, 350-, or 400-CID V8. The 400 was available in various ram-air configurations, including the fabled Ram Air IV package of '69—'70. As discussed in Chapter 4, the Ram Air IV was the epitome of performance development of the 400-cube engine. The option was as popular in Firebirds as in GTOs. It was an engine that oozed image: The name, the underhood view, the raspy roar when you wound out in the gears. Jim Wangers and the Pontiac crew knew how to merchandise a total experience.

Strangely enough, some of the most-exciting Firebird engine options came after the supercar era started to wind down in the early '70s. Pontiac engineers tried especially hard to keep performance alive after the fuel-economy/emission-control handwriting was on the wall. There was the 455-CID *H.O.* engine of the '71—'72 period, and the unbelievable Super Duty 455 of '73—'74. I'll detail some of these developments in Chapter 10. They had an important place in supercar evolution.

An important milestone in Firebird history was the introduction of the *Trans-Am* series in 1969. These Trans-Ams had nothing special under the skin that couldn't be had in other Firebirds. But their bold, flamboyant styling attracted a mass market of young blue-collar males that no one dreamed was out there.

Pontiac sold upwards of 100,000 Trans-Ams in '78 and '79, making huge profits that had the industry buzzing. In

Pontiac introduced their Firebird ponycar in late '67, shortly after the Camaro. Convertible bodies were especially popular in the Firebird line. Photo courtesy of Pontiac.

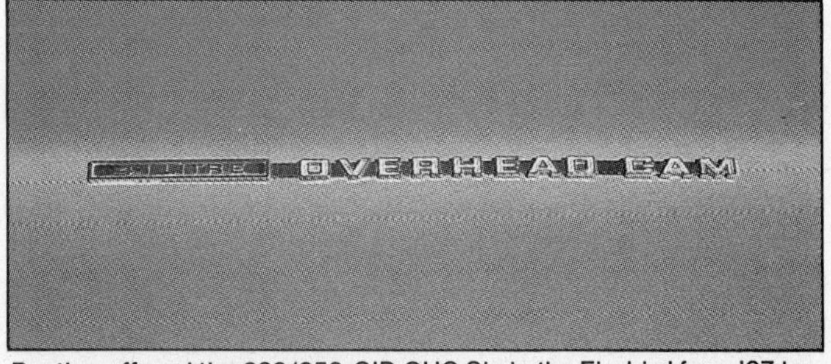

Pontiac offered the 230/250-CID OHC Six in the Firebird from '67 to '69. Photo by Ron Sessions.

fact, they say the $10-per-car royalty paid to the Sports Car Club of America for use of the *Trans-Am* name kept that organization afloat during some lean years in the '70s.

No other company was able to come near to duplicating the sales success of the Trans-Am Firebirds—though there were several attempts to copy it. Chevrolet tried the fender-flare/graphics bit with the Z/28 Camaro in the mid-'70s, but were initially outsold by an incredible 5-to-1 margin. That wild-bird decal on the Trans-Am hood had some *special* magic.

I have to credit Pontiac's head stylist, John Schinella, for a lot of the Trans-Am's magnetic styling. He had an uncanny ability to make a statement with lines and shapes. He was to Pontiac styling what Jim Wangers had been to Pontiac merchandising—an idea man.

Mercury Cougar—While I'm making comparisons, I could also say the Mercury Cougar was to the Mustang what the Pontiac Firebird was to the Camaro. Basically the same car, but a little fancier and a little heavier. In the case of the Cougar, the wheelbase was stretched 3 in. and the weight was up as much as 200 lb. And because practically the same lineup of engines was offered in the Cougar as in the Mustang, performance was inevitably down a bit.

Nevertheless, the original '67—'68 Cougar was a suprisingly clean design. Like the original Mustang, it was one of those rare Detroit cars that didn't look like it was designed by a committee.

As the story goes, Henry Ford II originally wanted the Mustang to be a mini-Thunderbird, decked out with luxury features. It's history that Iacocca prevailed, and the Mustang became a low-priced sporty car. With the Cougar, though, Henry got a second chance to go upscale in the ponycar segment. Thunderbird touches were everywhere: vacuum-operated hideaway headlamps, sequential rear turn-signal lamps, voided *compliance* bushings in the low-rate suspension and even a standard V8 engine. No Cougar strippos would go out the door.

Though the Cougar theme was one of personal luxury, Lincoln/Mercury Division was allowed to share the high-performance packages developed for the Mustang. Most notable were the '69—'70 428 CJ and '70 Boss 302 engines. It's interesting how Mercury engineers managed to mess up a great Boss 302 engine in their 1970 *Eliminator* series by fiddling with the air cleaner and exhaust systems to reduce noise. Of course, the extra 200 lb didn't help. Compare acceleration figures from CAR LIFE road tests:

	Boss Mustang	Cougar Eliminator
0—30 mph	2.9 sec	3.0 sec
0—60 mph	6.9	7.6
1/4-mile e.t.	14.8 @ 96 mph	15.8 @ 90 mph

Almost 1-1/2 million Firebirds were sold from '67 through '81. Photo courtesy of Pontiac.

Cougars of '67—'68 had exceptionally clean lines. This '68 model was available mid-year with the 428 CJ V8 and ram air. Photo courtesy of Ford Motor Co.

Mercury division used the Boss 302 engine to make the Cougar Eliminator in 1969. But excessive muffling of intake and exhaust noise took a toll on performance. Photo courtesy of Ford Motor Co.

When AMC signed Mark Donohue to drive a Javelin in the Trans-Am series, they celebrated the event by making a limited run of Mark Donohue Javelins for the street. This is a '70 390 SST model. Photo courtesy of John A. Conde Collection.

Needless to say, Cougars weren't popular street-racing hardware in the supercar era. They were, however, nice cars—forerunners of a whole flock of mid-size personal luxury cars that gained popularity in the '70s and '80s.

AMC Javelin/AMX—Smallest and least-wealthy of the U.S. auto companies, American Motors often required engineers to make one set of tools do two jobs. They managed this with interesting results on their 1968 entry in the ponycar market.

The basic design was known as the *Javelin,* with a wheelbase of 109 in. and weight comparable to the other ponycars of the period. AMC stylist, Dick Teague, came up with very clean, balanced lines for the car. Early Javelins were devoid of the fake hood scoops and brake-cooling slots prevalent on many Detroit ponycars of the period.

The cars had several nice touches. Recessed door handles, inside and out, were not only good from a safety standpoint, but also gave a neat appearance. Recessed instruments were set into a unique-design ABS-plastic intrument panel. And the SST model was the first ponycar to come standard with reclining bucket seats.

As with most ponycars, the Javelin was available with a wide selection of engines. In addition to the standard 232-CID six, early Javelins could be had with 290-, 343-, and 390-CID V8s. The 315-HP, 390 V8 was a good performer with a big Holley 4-barrel, 10.2:1 compression, forged-steel crankshaft and 2-1/4-in. dual exhaust system.

A few months after the Javelin was launched, AMC proceeded to whack 12 in. from the wheelbase of the car, just behind the door, and came up with the lithe and lively AMX two-seat sports coupe. Weight was reduced 250 lb. By putting their 390-CID 4-barrel V8 engine in the car, they achieved drag-strip times in the 14-second bracket. The AMX was definitely a competitive supercar of the period—and really quite aggressive looking.

It's a pity that only some 19,000 buyers could be found for the AMX over a period of three years. The two-seat AMX was phased out after the 1970 model run.

AMX coupes were high-performance ponycars in every sense of the term. With short 97-in. wheelbase, light and strong 390-CID engines, they were competitive against B-Production Corvettes in SCCA road racing. Photo by Larry Mitchell.

When AMC introduced the AMX, they rounded up land-speed-record man, Craig Breedlove, for a publicity exercise. Breedlove took the AMX out to North Texas and proceeded to set 106 national and world records with the car. Photo by Larry Mitchell.

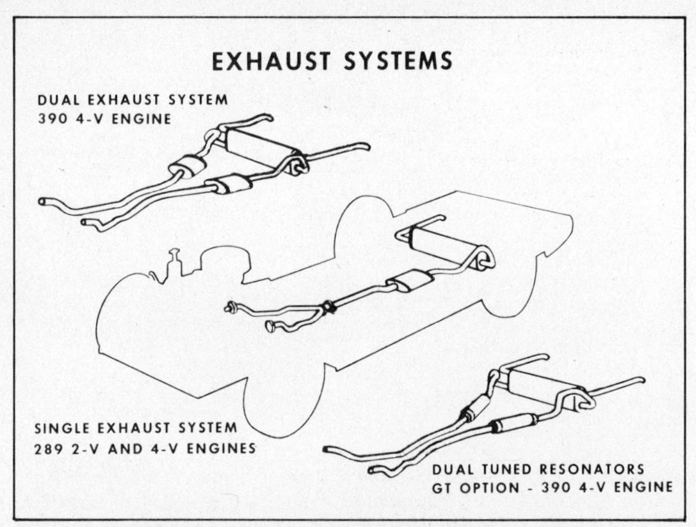

Ponycar performance was often hampered by inferior exhaust systems that used a single crossflow muffler behind the rear axle. This Ford Mustang system was more restrictive than the usual dual undercar system. Drawing courtesy of Ford Motor Co.

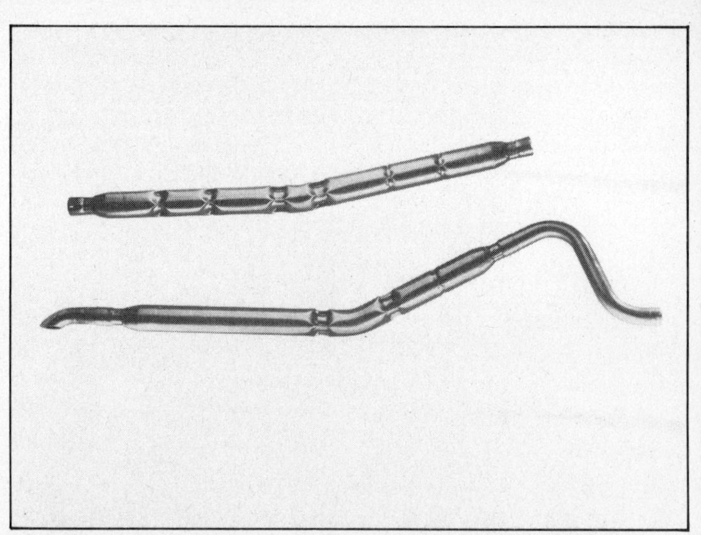

Chevrolet offered the unique Walker *chambered-pipe* muffling system as an option on some high-performance Camaros. But it was so noisy that the option was soon withdrawn.

In '71, the Javelin underwent a major redesign. Like other ponycars of the early '70s, it came off longer, wider and heavier. Performance was down, even with the big 401-CID V8. AMC kept the AMX name, but only as a model option on the 2+2 Javelin body.

A SPECIAL PERFORMANCE PROBLEM

Before discussing actual road-test figures on the ponycars, consider a special design problem inherent in many of these cars. There simply wasn't room for efficient exhaust systems. The early ponycars were relatively small cars. They were low, and the rearward engine position and short rear deck meant very little space under the floorpan. Due to ground-clearance requirements, big-diameter exhaust plumbing and large, low-restriction mufflers wouldn't fit. This was also true of other long-hood, short-deck cars of the period: Olds Toronado, Buick Riviera, and so on.

The solution was fairly simple. Put a large crossflow muffler parallel to and behind the rear axle, with one or two medium-diameter headpipes routed to it alongside the drive shaft. Small resonators could be placed just ahead of the axle, or in the dual-outlet pipes from the muffler. But the main muffling job was done by the big crossflow muffler behind the axle.

This design proved to be a good, durable exhaust system for utility passenger-car use—*but not so good for a high-performance car.* The single crossflow muffler was too restrictive—even with a minimum of internal baffling—and the medium-diameter pipe going back gave too much flow loss. The small resonators used by some companies were no problem. But the engineers wished they could use those for the mufflers—and end the plumbing right there!

It's unfortunate that most of the popular ponycars were hampered by this type of exhaust system. Notable exceptions were the '70—'74 Barracuda/Challenger, the '64—'66 High Performance Mustang, the upsized '71—'73 Mustang/Cougar and the Javelin/AMX. Also some of the hand-built specials, like the Boss 429 Mustang, used small undercar mufflers that were not strictly legal from a noise standpoint.

Many ponycar owners solved the problem by throwing out the factory system, and having a muffler shop fabricate a custom system, often using the smaller turbocharged-Corvair mufflers. These would add as much as 40 HP. The price: more noise and less ground clearance!

MEASURING 'EM UP

Now let's look at those CAR LIFE road tests on the standard and option-engined ponycars.

At the beginning of this chapter, I mentioned that the ponycars had the *potential* for better acceleration performance than the GTO-class supercars—simply because of 200-lb-or-so less weight. But if you look back at the performance figures for these cars in Chapter 4, you will see clearly that the ponycars did *not* have any better performance with a given engine combination. A perfect example is the L78 Chevelle vs. the L78 Camaro:

	Chevelle	Camaro
0—30 mph	3.0 sec	2.6 sec
0—60	6.6	6.8
1/4-mile e.t.	14.8 @ 99 mph	14.7 @ 99 mph
Estimated net HP	370	330

The key is obvious: Camaro's lower weight was offset by its restrictive exhaust system. Enough said.

THE SHELBY PHENOMENON

Mustang lovers will always be grateful that Carroll Shelby was not willing to accept the Mustang's performance and handling the way it came from Dearborn.

The wealthy Shelby, ex-race driver and master promoter, was able to work out a plan with Ford to modify Mustangs for high-performance street use. The standard High

Car	Transmission	Axle ratio	Adv. HP	0—30 mph	0—60 mph	1/4-mile e.t. @ MPH	Est. Net HP
'65 Mustang HP 289	M4	3.89:1	271	3.1 sec	8.3 sec	15.9 @ 85	190
'65 Barracuda Formula S 273	M4	3.23:1	235	4.0	8.2	15.9 @ 85	190
'67 Mustang 390 GT	A3	3.25:1	335	3.4	7.8	15.5 @ 91	250
'67 Camaro SS 350	M4	3.55:1	295	2.9	7.8	15.8 @ 89	230
'68 AMC AMX 390	A3	3.54:1	315	3.1	7.2	14.6 @ 96	290
'68 Barracuda 340 S	A3	3.23:1	275	3.0	7.0	14.9 @ 95	290
'68 Cougar XR-7 427	A3	3.50:1	390	2.9	7.1	15.1 @ 93	280
'68 Mustang 428 CJ	M4	3.50:1	335	2.5	6.9	14.6 @ 99	320
'69 Camaro SS 396 L78	M4	3.73:1	375	2.6	6.8	14.7 @ 99	330
'70 Javelin 390	A3	3.15:1	325	3.5	7.6	15.1 @ 91	250
'70 Challenger 440 Six Pack	M4	3.54:1	390	3.1	7.1	14.6 @ 98	360
'70 Plymouth 'Cuda 340	A3	3.54:1	275	2.9	7.5	15.0 @ 95	290

Performance 289 Mustangs were shipped to Shelby's shops in Venice, California. There his craftsmen more or less rebuilt the cars to semi-race specifications: Some were rebuilt to full-race specs. The cars were painted a distinctive blue and white, with special identifying trim, exclusive wheels, and so on. Shelbys were then distributed through selected Ford dealers around the country. The treatment didn't come cheap, even in mid-'60s dollars. Shelbys were routinely priced about $2200 more than garden-variety Mustangs.

Shelby's most-popular effort was the '65—'66 G.T.350 Mustang. This was based on the HP289 engine with manual four-speed transmission and 3.89:1 final drive. The engine was further hopped up with tubular-steel exhaust headers, special high-lift cam, aluminum high-riser intake manifold and a 600-cfm Holley carb with center fuel inlet.

Gross-horsepower rating was boosted 35 HP to 306 HP at 6000 rpm. Maximum revs were about 6800. A baffled, 6-qt cast-aluminum oil sump was used to prevent oil starvation in the corners. But there was much more to a Shelby Mustang than a high-performance engine and drive train. This was no mere straight-line, quarter-mile sled.

The suspension was basically High Performance Mustang. However, to improve cornering, Shelby moved the inner pivot of the upper control arms down 1 in. to give more negative camber at jounce. This also lowered the

Early Shelby Mustangs used the HP289 engine and distinctive blue striping on white. Other beneath-the-skin mods made Shelbys almost suitable for all-out racing in standard form. Very impressive performance and handling. Photo by Rick Kopec.

While early Shelbys were no-nonsense machines, a full complement of options was available. Here's the optional taillight treatment. Photo by Ron Sessions.

Cobra snake became the trademark of Shelby-modified Mustangs. Photo by Ron Sessions.

Ford HP289 engine was fitted with a high-riser aluminum manifold, 600-cfm Holley carb, and high-performance camshaft when used in the G.T.350 Shelby Mustangs. At a rated 306 HP, the Shelby version developed about 35-HP more than Ford's regular HP289.

All early G.T.350s were equipped with functional rear-brake cooling ducts. Photo by Tom Monroe.

front end about 3/4 in. Relocating the control-arm pivot effectively raised the front roll center—usually not a good thing. But in the G.T.350's case, it improved handling.

A 1-in.-diameter front anti-sway bar replaced the standard 3/4-in. bar. Traction Masters were added to the rear axle to control wheel hop. Koni adjustable shocks were fitted all around. And a reinforcing bar was installed between the front-spring towers to keep them from deflecting inward. Brakes were upgraded with harder disc-brake pads at the front, and segmented metallic linings in the rear drums. The rear brakes were cooled by honest-to-goodness functional scoops in the quarter panels. Goodyear 7.75 X 15 *Blue Dot* tires, rated for 130 mph, were run on 6-in.-wide aluminum or Magnum 500 steel wheels.

Weight wasn't overlooked either. A fiberglass hood, lightweight tubular-steel exhaust headers and removing the rear seats cut curb weight 150 lb. The battery was moved to the trunk to improve front-to-rear weight distribution.

There were also a few internal niceties: a big tachometer and oil-pressure gage in a special instrument-panel pod, sports steering wheel, identifying emblems on the dash and finned aluminum valve covers. All together, you'd pay an extra $20,000 for modifications this extensive today.

And Shelby wasn't just spinning his wheels, trying to make a silk purse out of what some thought was a sow's ear. The G.T.350 Mustang that came out of his back door was superior to Chevrolet's new Corvette Sting Ray in many ways. In fact, a G.T.350 won the SCCA B/Production road-racing championship in 1966 against tough small-block Corvette competiton. And the street Shelbys with Goodyear Blue Dot tires could corner with some exotic European sports cars. This was one of the most-successful conversions of a standard Detroit car of all time.

It's interesting to compare CAR LIFE acceleration times for a G.T.350 against those for a standard HP289 Mustang:

	HP Mustang	Shelby
0—30 mph	3.1 sec	2.4 sec
0—60	8.3	6.8
1/4-mile e.t.	15.9 @ 85 mph	14.7 @ 92 mph

While these straight-line elapsed times do not reflect the Shelby's overall performance advantage, they are nevertheless interesting. I credit the Shelby's advantage to its engine modifications: headers, camshaft, improved manifold and carburetor—plus the 150-lb weight reduction. It's basic physics that the Shelby enjoyed a better power-to-weight ratio.

In fact, my estimate of the G.T.350's true net-HP output is 220, or abut 30-HP more than the 190 HP estimated for the standard HP289. This agrees closely with Shelby's claim of 35-HP more gross output.

A Ford Mk. II race car leads a G.T.350 Shelby Mustang through the curves at Daytona in 1965. Shelby Mustangs gave Corvettes fits on the road courses.

A '68 G.T.500. 1967 and early-'68 models were available with the optional 427 V8. Photo by Steve Christ.

1969 Shelby G.T. 500. After 1967, Shelby relinquished production of his Mustangs to Ford. Photo by Tom Monroe.

Ford-built Shelbys didn't have the special under-the-skin features of the early Shelby-built cars. Only the name was the same. Photo by Tom Monroe.

Ford-built G.T.500 came standard with a 428-CID V8. This is a '69 model with the ram-air equipped 428 Cobra Jet. Photo by Tom Monroe.

Needless to say, Shelby Mustangs are priceless collectors' items today. Production was only about 2900 units in the '65—'66 period.

Did you know that Shelby produced a short run of '66 *G.T.350-Hs* for Hertz Rent-A-Car? A small fleet of Shelbys was pressed into daily service at airports and other rental-car outlets. Imagine being able to rent a first-class supercar today! Insurance underwriters would have a field day! By the way, the "Rent-A-Shelbys" were equipped with the milder 271-HP 289 and the upgraded C-4 automatic transmission.

In 1967, a G.T.500 model was added. A 428 V8 was standard and the more-powerful 427 was optional. But in mid-'68, the 427 was dropped.

The resulting cars were made considerably fancier, smoother, quieter—and *heavier.* That was the beginning of the end for the Shelby legend. The following year, Ford took over the production of the cars in a small Ionia, Michigan plant they obtained from A. O. Smith. Shelbys were dolled-up Mustangs after that.

Get an early Shelby if you want the real McCoy.

Details of Shelby Mustang 428 CJ. Photo courtesy of Ford Motor Co.

Plymouth Barracudas and Dodge Darts were major contenders in the SCCA Trans-Am races of 1966. Mustangs and Camaros dominated later. It was always an exciting race series.

Z/28 emblem distinguished this Camaro from other small-block contenders in the ponycar crowd. Photo by Ron Sessions.

TRANS-AM PONYCARS

The mushrooming popularity of NASCAR Grand National racing in the early '60s, with thundering full-size stock cars fender to fender on 170-mph banked speedways, spawned other types of stock-car racing. The Sports Car Club of America (SCCA) decided to promote their sedan road-racing classes. By drawing their rules from Appendix J of the International Sporting Code, they were able to attract some equipment from Europe. Some of the successful U.S. teams were later encouraged to race overseas.

The most-important development came in 1966. This was when SCCA officials combined FIA Group-II rules with a maximum wheelbase limit of 116 in. This became the famous *Trans-America Championship* series—*Trans-Am* for short. The series was aimed primarily at American compacts and ponycars.

It was well known that Chevrolet's Camaro would be introduced in the fall of '66 to challenge Ford's Mustang. So SCCA officials figured they could mix a little old Ford/Chevy rivalry with some European flair. To keep things from getting out of hand, a maximum displacement limit of 5 liters (305 cubic inches) was specified. The 116-in.-wheelbase limit kept out the big cars. A minimum production requirement of 1000 units would supposedly keep out the hand-built race cars. Manufacturers were required to file homologation papers with the FIA in Paris to make specific engine/body/chassis combinations eligible. These specs guided tech inspectors at the tracks.

Actually, the International Group-II rules allowed plenty of elbow room for performance development. In the engine compartment, you had to maintain stock bore and stroke, stock cylinder heads and a factory intake manifold. But aftermarket pistons, camshaft, valve springs, carburetor, exhaust headers, and so on could be used. Engines were an interesting scramble of factory and hot-rod parts.

Extensive suspension, steering and brake mods were permitted to ensure safe high-speed handling. Special racing tires could be used—but they had to fit on stock-dimension wheels. Bodies had to remain stock. You could not even bubble the hood to clear an exotic carburetion system.

Nothing too wild happened that first season. Bob Tullius won the first race at Sebring in March, 1966. driving a Dodge Dart with a 273-CID small-block V8. But the 289-CID Mustangs came on strong later in the season—helped by Shelby's high-riser *factory* manifold. Purses were increasing fast, even after the first few races. Chrysler was leading the manufacturer's-points championship going into the last race, but Ford won that race and the championship in an exciting finish. Things looked great for the coming '67 season.

Camaro Z/28—It was at this point that Chevrolet stepped in and made a whole new ball game of Trans-Am racing. That is, up to that time no factory models were aimed specifically at Trans-Am-class racing. Teams picked a *catalog model* that fit the wheelbase limit, with an engine close to, but under the 5-liter-displacement limit. The car was then prepared to race with the best combination of factory and aftermarket equipment that could be assembled. Mustangs won in '66 because they had a little-better combination than the Barracudas and Darts.

But Chevrolet wanted to promote the new Camaro through Trans-Am racing. Vince Piggins' famous *Product Promotion* group was behind it. They wanted a combination that could win consistently—with enough legal factory equipment to give amateur racers a 50-HP advantage over the 289 Mustangs. Furthermore, Chevy was willing to build the necessary 1000 units of a special Camaro "Trans-Am" model if that was what it took to do the job: They meant business.

What emerged in the summer of '67 was the famous Camaro Z/28. Its target was Trans-Am racing.

To get engine displacement closer to the 5-liter limit, they combined a 283 crankshaft with the 327 block. From high-performance 327-Corvette engines, they borrowed the big-valve cylinder heads, forged pistons, heavy-duty rods, and the wild *30-30* solid-lifter cam. The bottom end was further beefed with four-bolt main-bearing caps,

In '68 a rear spoiler and hood/deck striping provided additional Z/28 identification. Photo by Ron Sessions.

Chevrolet developed a new 302-CID high-performance engine for the Z/28 Camaro, using some off-the-shelf goodies and some new ones. Net output was said to be 290 HP at 5800 rpm. Photo courtesy of Chevrolet.

steel crank, floating piston pins, windage tray, deep pan and a high-volume oil pump.

On top, Chevy engineers installed a newly designed aluminum high-riser intake manifold—and topped it with a big 780-cfm Holley 4-barrel with center-hung floats for high-G cornering. Chevy stopped short of putting tubular-steel exhaust headers on the engine—there was a set in the trunk, ready to bolt on, if you wanted to race your Z/28!

The chassis wasn't forgotten either. The standard setup included a carefully calibrated combination of high-rate springs and stiff-valved shocks, harder brake linings, 7-in.-wide wheels, 70-profile tires, close-ratio four-speed transmission, and 3.73:1 Positraction rear end. Cornering and general handling were outstanding for the day. Serious racers could go a lot further with chassis refinement using equipment right out of the factory parts book. This included yet higher-rate springs and stiffer shocks, heavy-duty suspension arms, frame reinforcements and faster steering. There was even a kit to adapt the Corvette's four-wheel disc brakes to the Camaro chassis.

Chevrolet was clever with Z/28 styling, too. Nothing too bold—but there was no doubt it was a very special Camaro. Z/28s had two broad stripes the length of the hood, roof and rear deck, a modest rear spoiler lip, understated little emblems on the front fenders, and sporty rally wheels that had the deep-dish look because of the wide rims. Neat. And definite.

At a total price around $3300, it was bound to sell. First-year sales were only around 600 units because of the late start and limited production capacity. But 7200 units moved out in 1968; Chevy dealers began to notice. Sales of 19,000 Z/28s the following year astonished everyone. Market analysts read it as evidence that a car's image had more influence in the youth market than bread-and-butter factors like smoothness, driveability and response. Those early 302-CID Z/28s were anything but flexible street machines. They were thinly disguised race cars. But Chevrolet sold them by the thousands.

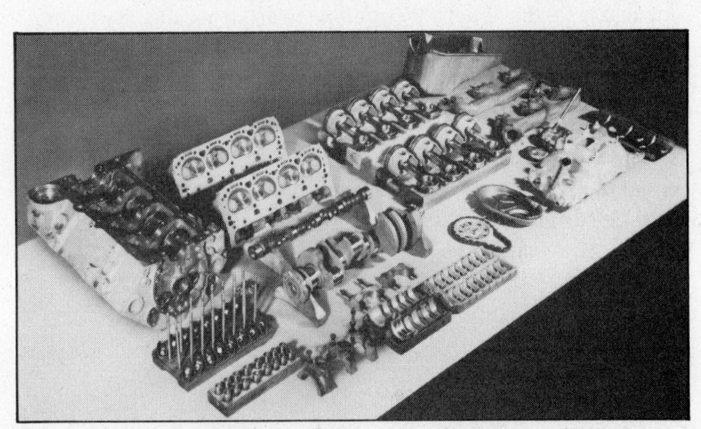
Special parts used on the 302 Z/28 engine. As compared with the standard small-block Chevy, the 302 was practically a complete ground-up conversion. Photo courtesy of Chevrolet.

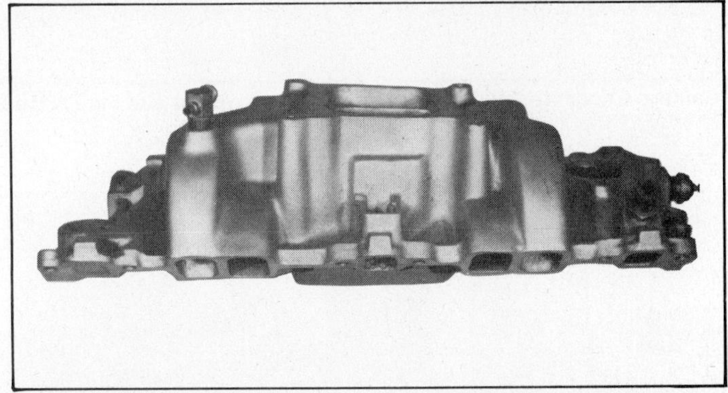
Two-plane aluminum high-riser manifold designed for the 302 Z/28 was the best induction system yet for the small-block Chevy. Passages were very large and smooth. Photo by Mike Crawford.

Sales snowballed because the cars were earning a performance reputation on the race tracks. Mustang won the Trans-Am points championship in 1967, but the Z/28 didn't appear until mid-season. The Sunoco-sponsored Penske Camaro, with Mark Donohue as the top driver, won 18 out of 25 Trans-Am races in 1968 and '69. Like the

Though early Camaro Z/28 engine performance was virtually unsuitable for the street, these cars were an immediate hit with the street crowd. Photo by David Gooley.

1969 Z/28 with the rare Rally Sport front end. Photo by David Gooley.

Camaro's major restyling in 1970 added some weight. But the Z/28 engine was upped from 302 to 350 CID to offset it. Chassis development improved handling. Photo courtesy of Chevrolet.

Mark Donohue's Sunoco-sponsored, Penske/Hilton Racing '68 Z/28 Camaro. Photo by Ron Fournier.

Rear view of Donohue's '68 Z/28. Notice large-diameter side pipes, full roll cage and fuel cell in trunk area. Photo by Ron Fournier.

Chevy engineers developed an unusual cross-ram manifold, with two 600-cfm Holley carbs, for Trans-Am racing with the Z/28. It gave a very fat torque curve from 4000-rpm up. Photo courtesy of Chevrolet.

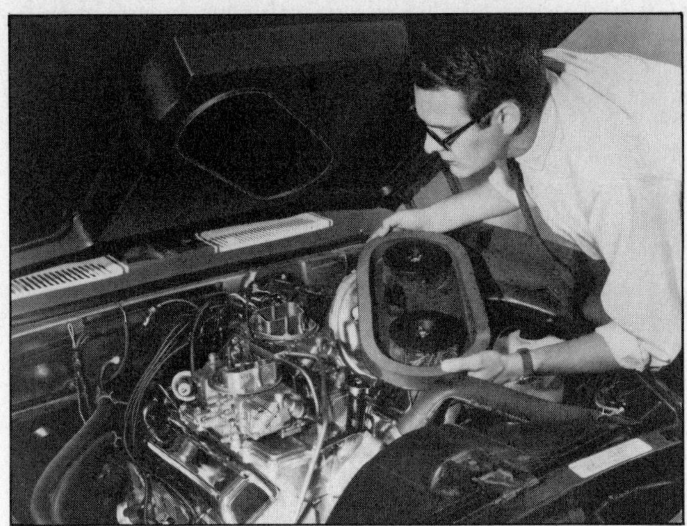

Chevrolet intended to offer the cross-ram carburetion system as an option on the Z/28—and developed air cleaners and a cold-air hood. The idea was dropped before production started. Photo courtesy of Chevrolet.

Boss 302 Mustang was an exciting car. Ford used side striping, spoilers and wheels for identification, being sure not to copy the Z/28! Photo courtesy of Ford Motor Co.

Ford engineers developed exotic equipment for the Boss 302 Trans-Am racing program. This display shows the dual Holley 4500 carburetion system that was used briefly in 1969. These 1100-cfm "elephant" 4-barrels were later banned because they were not production carburetors. Photo courtesy of Ford Motor Co.

old NASCAR stock-car days, Z/28 sales seemed to parallel the number of victories on the SCCA road courses.

It's interesting, incidentally, that an important piece of factory equipment that helped earn these victories was never offered as an option on street Z/28s. This was the unique dual 4-barrel cross-ram induction system. The manifold was like a large tub, with long criss-crossing ram passages cast in the bottom, and two 600-cfm Holleys set diagonally on the top lid. The ram effect gave a fat torque curve from 4000 to 6500 rpm. It was great for accelerating out of slow corners. The system certainly had lots of eye appeal. But Chevy engineers discouraged it for street use. Only a handful of these manifolds were made—and needless to say, factory originals are pretty scarce today.

MORE TRANS-AM PONYCARS

Boss 302 Mustang—It could be predicted that Chevrolet's quick success with their "Trans-Am" Camaro would attract copiers among the other ponycar manufacturers. And sure enough, in 1969, Ford introduced the Boss 302 Mustang. It had all the established features of the Z/28: special identifying body trim, heavy-duty suspension, wide wheels and low-profile tires, heavy-duty brakes, four-speed transmission, numerically high 3.91:1 final-drive ratio, and more.

But in the case of the Boss 302, the *engine* combination was especially interesting. Ford made a fire-breather out of their small-block wedge V8.

Ford was in the process of tooling up a new 351-CID

Ford offered the Boss 302 Mustang in '69 as an answer to the Z/28 Camaro. Engine had finned valve covers and the usual open-element air cleaner. A cold-air system was standard in 1970. An *export* brace between the fire wall and shock towers and a *malibu* bar between the shock towers added necessary rigidity to the front end.

Ford used big-port *Cleveland*-type cylinder heads for the Boss 302, because they had the same bolt pattern and bore-center spacing as the *Windsor* engine. But the ports were almost too big for 302 cubic inches.

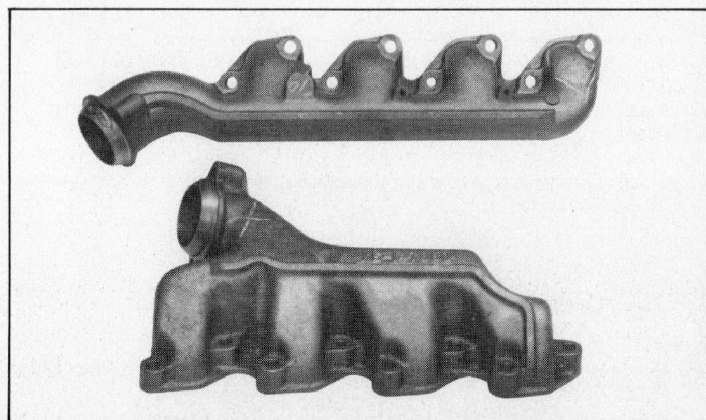

Ford also designed neat low-restriction exhaust manifolds for the Boss 302 Mustang. Right side had smaller passages because of chassis-clearance problems. This was a problem on many performance ponycars.

335-series high-performance engine to be built in their Cleveland engine plant. The all-new engine featured staggered valves, *semi-hemi* combustion chambers and very large ports—nice stuff, not altogether unlike the Mark IV big-block Chevy. As a matter of economics, bore spacing and cylinder-head bolt pattern for the new *Cleveland* engine were the same as the older small-block wedge being made in Windsor, Canada. The benefit was that some of the machining could be switched back and forth.

But think for a minute what horizons this opened up for that mundane little 302-CID *Windsor* V8. With bore spacing and bolt patterns shared with the Cleveland engine, it was a natural to bolt heads similar to the big-port *351C* heads on the 302-Windsor block. The swap gave the Boss 302 Trans-Am engine more breathing capacity than it could use. And it was done without any big investment in special tooling. As in the case of the Z/28, the Boss 302 Mustang engine was made primarily by scrambling existing production parts.

New parts included an excellent aluminum high-riser intake manifold, 780-cfm Holley carb, a new 290° solid-lifter cam and high-rate valve springs. Down below, the Boss 302 had four-bolt main caps, 11:1 forged pistons, forged-steel crankshaft, heavy-duty rods from the HP289 engine, and the usual deep oil pan, windage tray and high-capacity oil pump. Ford also spent a few extra bucks on special low-restriction exhaust manifolds to feed the dual undercar exhaust system, where Chevy used stock manifolds. Chevy believed that most owners would switch to tubular headers immediately. But few Z/28 owners did and Ford's clean manifolds on the Boss 302 gave it a horsepower advantage above 6000 rpm.

It's also interesting, in comparing the Z/28 and Boss 302, that both companies rated their engines at 290 HP at 5800 rpm. And both said this was *net* power, not gross power. Subsequent acceleration tests suggested they were not exaggerating. Both were very strong street engines, pulling nearly 1 horsepower-per-cubic inch in showroom trim with street exhaust systems.

Ford Engineering developed a lot of special performance parts for their Trans-Am racing effort that were never offered on street Mustangs. These included such engine exotica as an *in-line* 4-barrel Autolite carburetor, specially forged, super-duty connecting rods from the four-cam Indianapolis engine, needle-bearing rocker arms, triple valve springs, titanium valves, and so on. Chassis goodies included: adjustable rear sway bar, full-floating rear axle, four-wheel disc brakes and rear-suspension track bar.

Ford was spending millions on racing in those days, but didn't get nearly the results-per-dollar that Chevrolet did.

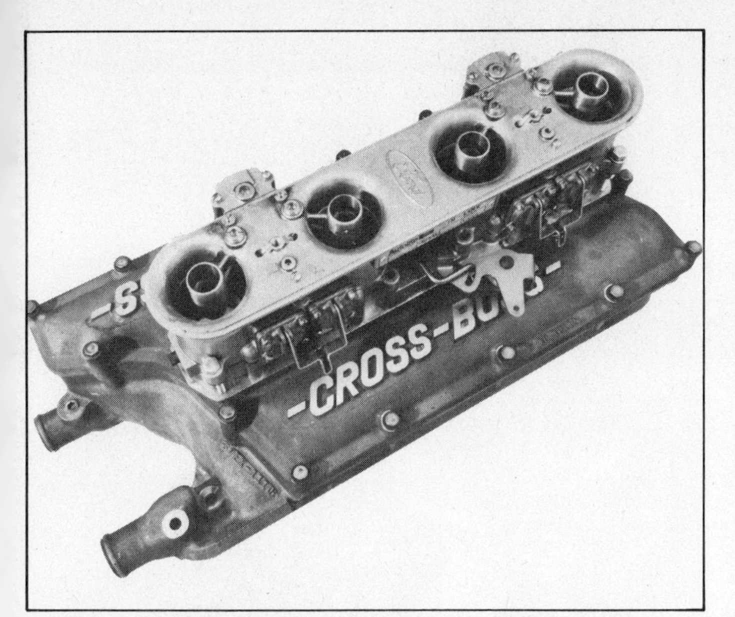

Another Ford attempt at carburetion exotica: An Autolite *in-line* 4-barrel carburetor on a cross-ram manifold for the Boss 302 heads. Just about the ultimate in 4-barrel carburetion, the Cross Boss had 1200-cfm capacity.

Parnelli Jones, ex-Indianapolis ace, was the top Mustang driver in Trans-Am racing in the late '60s. It was a fast, wild, fender-banging style of racing that attracted big crowds.

Ford nosed out the Penske Camaros for the Trans-Am championship in 1970. But Henry II shut down all racing development overnight: 1970 was the last season the Detroit companies really contested the series.

AAR 'Cuda/Challenger T/A—Chrysler was also in the Trans-Am picture in 1970. Taking advantage of a new rule that allowed larger-displacement engines in *production versions* of the Trans-Am ponycars, Dodge and Plymouth used special four-bolt-main 340-CID Six-Pack engines in new Challenger T/A and AAR 'Cuda models. In addition to 6-barrel carburetion, these engines also featured an extra-long-duration cam, offset rocker arms and pushrods to allow wider head porting, and beefed valve gear. Interestingly enough, the heads came from the factory with stock-size ports. Racers wanting to take advantage of the offset rockers had to remove the heads and grind and match the ports to intake-manifold size!

Street versions would rev safely to 6500 rpm with hydraulic lifters. Power was also helped by a sophisticated air-induction system. The Dodge system featured a raised hood scoop designed in the wind tunnel to get full ram effect at about 80 mph. The Plymouth system was a little different: It used a NACA scoop to get ram air at high speeds.

These Chrysler ponycars were hampered by 200-lb more weight than the '70 Camaros and Mustangs. But magazine strip tests showed very competitive acceleration figures, indicating strong net output. And one extensive handling/cornering test showed good potential road-course performance—as you'll see later. Obviously Chrysler engineers did a lot with what they had.

Firebird Trans-Am—GM's Pontiac division aspired to Trans-Am immortality with a short-stroke 303-CID Firebird in 1969. A considerable amount of money was spent on special tunnel-port heads, forged cranks, ram manifolds, and the like. But management never committed to the necessary 1000-unit production program. The thing ended up with a couple of Trans-Am teams running 302-CID Chevy engines in Firebirds—on the basis that Firebirds built in Canada used Chevy engines.

Dan Gurney and his AAR team did fairly well in 1970 Trans-Am racing with the new 340 'Cuda.

With only halfhearted factory support, the Firebirds were barely competitive at the hands of Jerry Titus. Eventually the small stock of Pontiac Trans-Am parts were installed on a handful of 400-CID four-bolt-main blocks: These became the elusive *Ram Air V* engines that every Poncho fan has dreamed of finding. I doubt if more than three or four were ever installed in cars.

MEASURING UP THE TRANS-AM PONYCARS

My files contain reliable CAR LIFE road tests on several of the Trans-Am ponycars—all tested in strictly showroom trim, with no special equipment or tuning. The table on the following page shows how they compared.

Model	Transmission	Axle ratio	Adv. HP	0–30 mph	0–60 mph	1/4-mile e.t. @ mph	Est. Net HP
'69 Camaro Z/28 (302)	M4	4.10:1	290	3.5 sec	7.4 sec	15.1 @ 95	290
'69 Boss 302 Mustang	M4	3.91:1	290	2.9	6.9	14.8 @ 96	280
'70 Camaro Z/28 (350)	A3	4.10:1	360	2.9	6.5	14.5 @ 99	320
'70 AAR 'Cuda 340-6V	M4	3.55:1	290	—	—	14.5 @ 99	310

Original '69 Firebird Trans-Am copied the Z/28 hood and deck striping. And they're rare. Less than 700 were built that year. Photo courtesy of Pontiac.

Here it is, big as life: The mystical Pontiac Ram-Air V engine package. Actually a conglomeration of leftover Trans-Am racing parts—fitted to a 400-CID four-bolt-main block. None were sold to the public.

The above figures show that standard-production Trans-Am ponycars, with their small-block engines, could turn acceleration times about equal to the big-block supercars in the GTO class. The key was their lighter weight.

Also, it's obvious that these highly tuned small-block engines were putting out a lot more horsepower *per cubic inch* than the big-block street engines. Though based on ponycars, the Trans-Am derivatives weren't wheezing through restrictive intake systems. If you study their specs, you'll find much-better carburetion, camming, valve area, and so on.

Trans-Am engines like the Z/28 and Boss 302 were closer to all-out race engines than their counterparts in bread-and-butter supercars such as the GTO, Dodge R/T or Olds 4-4-2. And by the same token, they were less convenient to drive in city traffic. They tended to be rough and noisy, sluggish in low-end response, and peaky at the top end. Fuel economy was acceptable only because the engines were relatively small. Living with these cars required a lot of loving.

In fact, the early 302-CID Z/28s were a real problem on the street. The sluggish 0—30 and 0—60-mph acceleration times tell the story. With the close-ratio four-speed transmission, small cubes and long-duration 30-30 cam, there wasn't enough low-end torque to pull the skin off a grape. The engine didn't really start to pull until 5000 rpm. It was a horrible combination for street racing. Owners of those early 302 Z/28s would try to get you to race from a 20-mph rolling start! It wasn't until Chevrolet switched to the 350-type block and a shorter-duration cam in 1970 that the Z/28 became a viable street racer.

Mark Donohue with his Penske/Sunoco Javelin. American Motors sponsored a big racing program in 1970, but Ford won the Trans-Am crown that year. AMC finally won it in '71 and '72, after the other factory teams pulled out. Photo courtesy of the John A. Conde Collection.

As far as that goes, the same driveability problems applied to the Boss 302 Mustang. But in this case, the problems stemmed from the engine's huge ports and valves. They were much too big to feed 302 cubes. At revs below 6000—6500 rpm, the low velocity of the air/fuel charge resulted in less torque than with the standard 302 V8.

The only thing that saved the street performance of the Boss 302 was the low final-drive ratio and wide-ratio four-speed. But those big-port 351C heads were never right for the 302 block. Even the professional Trans-Am racers, who routinely revved the engine to 8000 rpm, would often line the intake ports with epoxy to make them *smaller!* It

1970 marked the last year the Trans-Am series was hotly contested. Here, Tony DeLorenzo in an Owens-Corning-sponsored Camaro leads Ed Leslie and Mark Donohue going into a turn. Photo courtesy of Owens-Corning Inc.

TRACO developed American Motors V8 engines for several types of racing in the early '70s, including Trans Am, NASCAR, and sprints. The engine proved to have good potential.

gave them better mid-range torque off the turns, with practically no sacrifice at the top end. When you can do that to an engine, and make it perform better, something is radically wrong.

Perhaps the Dodge and Plymouth Trans-Am cars with the 340-CID Six-Pack engine had the best street combination of all.

HOW RACEWORTHY WERE THEY?

Straight-line acceleration tests don't tell much about how a car might perform on a twisty road course. Fortunately the editors of CAR LIFE magazine were just as interested in the raceworthiness of the factory Trans-Am offerings as they were in their street-racing potential. After all, many buyers of these cars wanted them primarily for their superior handling, cornering and braking—not their brute speed and acceleration. These were considered poor-man's Corvettes by many, and they were expected to perform as such.

Detroit's involvement in Trans-Am racing peaked in 1970. That year, the editors of CAR LIFE decided to try to measure the handling characteristics of the popular factory Trans-Am cars. It proved to be a challenging job. You can measure acceleration with a stopwatch and fifth wheel. But, back then it was hard to measure something as subjective as handling.

After several false starts, they came up with a three-phase test they felt would put equal emphasis on acceleration, braking, steady-state cornering and general steering response. All of these categories were considered vital ingredients in a car's speed potential on a typical road course.

The three tests: (1) Minimum braking distance from 80 mph, (2) Maximum average lateral *G-force* developed in steady-state cornering on a skidpad, and (3) Minimum time to cover an 0.8-mile slalom course from a standing start—which included tight turns, broad turns, short straightaways, hard braking and a short zig-zag stretch between pylons. This course was supposed to test everything that counted to the sporty driver.

The contesting showroom stock 1970 cars:

AMC Javelin, 360 CID
Camaro Z/28, 350 CID
Boss 302 Mustang
Plymouth AAR 'Cuda, 340 Six-Pack
Firebird Trans-Am, 400 CID Ram Air

Here are the numbers that came up:

BRAKING FROM 80 mph (in feet)	
Boss Mustang	308
Camaro Z/28	321
AAR 'Cuda	326
Firebird Trans-Am	332
Javelin	349

SKIDPAD G-force (average right and left)	
Boss Mustang	0.73
Camaro Z/28	0.72
AAR 'Cuda	0.70
Firebird Trans-Am	0.69
Javelin	0.61

LAP TIME FOR SLALOM (in minutes: seconds)	
AAR 'Cuda	0:59.4
Firebird Trans-Am	0:59.5
Camaro Z/28	1:00.0
Boss Mustang	1:02.5
Javelin	1:03.8

Figures tell the story. There's not much I can add. The Boss 302 Mustang had the best braking and skidpad figures—but couldn't get it together on the tight slalom course. Drivers mentioned that the lack of low-end torque coming off the slow corners hurt them. Firebird was helped by the opposite situation—smaller ports and big cubes. The Z/28 was just about in between.

What can I say about the Javelin? Certainly no slouch, it suffered from a lack of development—pronounced *bucks*—that the other carmakers could lavish on their cars.

'Cuda won by having it all together: Decent braking, decent cornering, and a good compromise between top-end power and mid-range torque.

Ponycars were fun!

1963—'64 Falcon Sprints were equipped with the 260-CID 2-barrel V8. Photo by Ron Sessions.

Falcon Sprints were true budget supercars. They came standard with plain-Jane bench seats and three-speed column-shift manual transmission. Nevertheless, styling was clean and snappy. Photo by Ron Sessions.

Junior Supercars

More go for less dough

DE-FATTING THE GTO

Within six months after Pontiac's GTO hit the showrooms in the fall of 1963, every Detroit sales executive knew where to make big money—in the youth-oriented image/performance market. Those first GTOs were backordered for months, and Pontiac dealers weren't discounting a nickel. This new type of car definitely filled a slot in the market not served by any other American or imported car. The potential looked almost limitless.

But there's a simple rule of merchandising economics that says—*reduce the unit price and broaden the potential market.* No one has ever argued with it. More people will buy a $10 item rather than a $100 luxury version of it.

What this meant to the budding supercar market was obvious. The most-popular-optioned GTO was a *medium-priced* car. It had a lot of luxury features such as bucket seats, center console and special instrument panel. It also had a big-displacement V8 engine. Though it enjoyed better fuel economy than the wild super stocks of the early '60s, it was still a *guzzler* in comparison with standard passenger cars. Actually, none of the GTO-type cars were *economy* cars in any sense of the word—initial price, trim level or fuel economy.

So Detroit sales people began to wonder, "Why not a *junior supercar?*" Maybe a lighter compact body rather than an intermediate. Fewer fancy frills. But still plenty of image and model identification. And, of course, a *small-displacement* engine with enough muscle to give good performance in the lighter-weight car—but small enough for better fuel economy. Lastly, increase market potential by holding the package price $500—1000 below the GTO-class cars. Not a bad formula.

By 1970, most of the companies making conventional supercars had a go at some type of budget combination that they hoped would appeal to a broader segment of the youth market. There were many interesting ideas.

Not all stuck to the formula of a small-block V8 in a compact car. A few slipped a big-block into the picture—but trimmed frills to the bone to keep weight and price down. Some used intermediate bodies, but stripped all excess weight.

Pontiac tried to squeeze into the market by putting an overhead camshaft on a bread-and butter six-cylinder block. It had enough extra goodies to give exciting acceleration and 6500-rpm revving ability.

It was a fascinating market. The cars didn't offer the all-out acceleration and speed of the classic big-block supercars. But they definitely had an important place in the evolution of high-performance cars in America.

Let's take a closer look at some specific combinations tested by CAR LIFE in the mid- and late '60s.

1964 Ford Falcon Sprint—This car couldn't be said to be inspired by the GTO, as it actually predated the "Goat." Introduced in '63, the Falcon Sprint was Ford's attempt to put a little zip in a three-year-old line of compact economy

Above: Falcon Sprints were popular with budget-performance buyers. This is a '63 model.

Chevrolet made an economy/performance breakthrough when they offered the 283 4-barrel V8 in the small 1965 Chevy II SS. In '66 they really stirred the performance ranks with the 350-HP 327. Photo courtesy of Chevrolet.

Chevrolet restyled the compact Chevy II in 1968, and renamed it *Nova*. Larger and heavier, it was available with small- and big-block engines up to 427 CID. This is the '68 SS 350. Photo courtesy of Chevrolet.

cars. Ford used their 260-CID 164-HP small-block 2-barrel V8 for power, which promised decent acceleration in a car weighing around 3000 lb. Bucket seats, a center console and special trim were added. Ford sold the two-door hardtops and convertibles for around $2800. Sprints were clean and snappy-looking little cars. Mercury introduced a similar Comet Caliente model at the same time.

CAR LIFE's acceleration figures on a 3100-lb Sprint convertible four-speed weren't all that great:

0—30 mph	3.9 sec
0—60 mph	12.1 sec
1/4-mile e.t.	18.0 sec @ 75 mph
Est. net HP	130

Observers wondered why Ford didn't use the high-performance 289 engine for the Sprint package. Simple: They were afraid the tiny 7-1/2-in., integral-carrier rear axle wouldn't hold up. Ford could have used a bigger axle, but its engineers were very busy with the new Mustang. The HP289 Sprint program was shoved to the back of the stove. Result: Practically no one remembers the Falcon Sprint!

1966 Chevy II with 350-HP 327 is still a winning combination at the drag strip. This one runs in SS/LA. Photo by Ron Sessions.

1966 Chevy II SS 327—Nothing too special about an *SS* or Super Sport model from Chevrolet. Basically a trim package, Super Sports were available in almost every car line. It was what was under the hood that made the difference.

Chevy first made the small-block V8 available in the

In the lightweight Tempest Sprint, the Pontiac Overhead-Cam Six delivered good fuel economy with 7000-rpm rev potential. Photo courtesy of Pontiac.

Pontiac put an overhead camshaft on the Chevy six-cylinder block in 1966 to make a budget performance engine. This high-performance version with 4-barrel carb was rated at 207 HP. Standard in the Tempest Sprint was a 1-barrel version rated at 167 HP. Photo courtesy of Pontiac.

compact Chevy II in '64. With 283 cubes and 2-barrel carburetion, performance was crisp but not spectacular. In 1965, the 327-CID 4-barrel version of the small-block engine was offered in the Chevy II. That was a performance breakthrough. But when the fabled 350-HP Corvette combination was released the following year, the CAR LIFE testers just had to get out their stopwatches.

Actually this 350-HP 327 was one of the sweetest and best-performing small-block Chevy engines of that era. Chevy engineers scrambled existing off-the-shelf parts in a brilliant way to achieve the high-end/low-end compromise they needed for a medium-weight car. They took big-valve heads and 11:1 pistons from the fuel-injection engine and added Carter AFB carburetion from the 300-HP engine. Chevy also designed a new hydraulic-lifter cam with 0.440-in. lift that could pull to 5800 rpm with standard lifters and valve springs. Yet, the engine had a smooth idle and strong low end. Tremendous engine.

With this engine, close-ratio four-speed and 3.31:1 gears, the 3100-lb Chevy II had near-supercar acceleration:

0—30 mph	2.6 sec
0—60 mph	7.2 sec
1/4-mile e.t.	15.1 sec @ 93 mph
Est. net HP	250

As an interesting aside, CAR LIFE later tested a '70 Nova SS with a 350-CID 4-barrel engine, and couldn't duplicate the 327-Chevy II's figures. It was thought that the extra cubes would just about offset the 200—300-lb weight increase of the later model. It didn't.

It required the 396-CID big-block engine to make those Novas of the late '60s and early '70s really move. But these Novas were by no means *junior supercars* in performance *or* price.

1966 Tempest Sprint Six—The Pontiac performance crew sneaked in the back door with the Tempest Sprint. The original idea behind an overhead camshaft on the 230-CID Chevy six-cylinder block was to add some excitement and technical novelty to the line of Tempest economy cars—which it did. But then Jim Wangers and John DeLorean and company got the idea of hopping up the combination to junior-supercar status. They even dared dream of producing a poor-man's Jaguar—based on the screeching rasp of a long-stroke six-cylinder turning more than 6000 rpm.

The engine had potential. The overhead cam eliminated the reciprocating mass of lifters and pushrods, making 6500 rpm possible without excessively stiff valve springs. The absence of pushrods allowed the engine to breathe through bigger ports and valves. The baseline version with standard 1-barrel carb boosted output 25 HP over the old pushrod six, from 140 to 165 HP. On the optional high-performance version, Pontiac pushed breathing even further by adapting the GM Quadrajet 4-barrel carb on a special ram manifold, plus adding a high-lift camshaft and a tuned, low-restriction exhaust manifold. A final touch was a healthy 10.5:1 compression ratio—and, voila, power rose from 165 to 207 HP at 5200 rpm.

It was truly a poor-man's Jaguar. And yet the use of rubber cog belt to drive the overhead camshaft, and unique hydraulic tensioners to absorb valve-train lash, made the engine as quiet and smooth as any conventional pushrod six.

CAR LIFE'S test of a 1966 Sprint with 3.55:1 final drive and the new wide-ratio Saginaw four-speed showed healthy acceleration—for a six:

0—30 mph	3.0 sec
0—60 mph	8.2 sec
1/4-mile e.t.	16.7 sec @ 83 mph
Est. net HP	190

Olds F-85 Jetfire of '62—'63 was an engineering showcase. Not only did the Jetfire represent Oldsmobile's first use of turbo-supercharging in a production car, but the all-aluminum 215-CID V8 achieved that magical figure of one advertised horsepower per cubic inch. Photo courtesy of Oldsmobile.

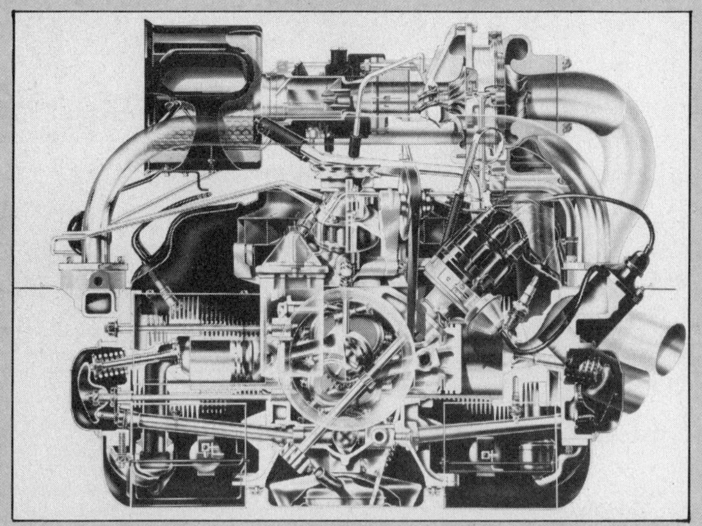

In '62-1/2—'64, the turbocharged 145-CID Corvair flat six developed an advertised 150 HP. 1965-'66 version had 164 CID and 180 HP! Photo courtesy of Chevrolet.

Dick Griffin of Lansing, Michigan gained a national reputation in the early '60s by getting supercar performance out of turbocharged Corvairs. Only relatively simple mods were required. Photo by Demmer.

GM TURNS TO THE TURBOCHARGER

American-supercar fans saw an interesting short-lived period in the early '60s when two GM divisions tried the turbo-supercharging trick. This was done to get *super-stock* performance out of compact economy cars. Midway into the 1962 model year, Chevrolet and Oldsmobile introduced turbocharged versions of the Corvair and F-85 Cutlass. Both installations used 5—6-psi boost pressure with moderate compression ratios to run on available premium fuels. Otherwise the two engines were entirely different.

Corvair used a 145-CID air-cooled flat six-cylinder engine with a Thompson TRW turbocharger drawing through a side-draft Carter YH carburetor. In the Olds Cutlass Jetfire was a fascinating little 215-CID all-

Turbocharged Corvairs were real sleepers with uncluttered lines. Only external identification was a small turbo logo on the engine-compartment lid. Photos by Ron Sessions.

Unlike the short-lived Jetfire, the Corvair turbo enjoyed a production run spanning a half decade. Photo by Ron Sessions.

Chevrolet didn't skimp on instrumentation for their budget Corvair supercar. This '65 Corsa was equipped with a cylinder-head-temperature gage and manifold-boost gage. Photo by Ron Sessions.

aluminum V8 that kept car weight well under 3000 lb. It was only 200-lb heavier than the air-cooled Corvair. On a basis of cubic-inches-per-pound, the blown Cutlass had more performance potential.

Of technical interest was the way the engines were protected from destructive detonation. Remember that turbocharging was in its infancy in those days and engineers were naturally skittish about unexpected field conditions. Chevrolet engineers protected their engine by purposely building in excessive intake and exhaust restrictions: The engine was choked to death above 4000 rpm. You could blow the engine by putting on a bigger carburetor and exhaust-turbine housing!

Olds engineers used an entirely different tack. They rigged up a system to inject a water/alcohol mixture into the compressor above 2-psi boost. This was very effective in suppressing detonation. And for added safety, an interlock system prevented full-throttle opening when the injection-fluid tank was empty. I can recall going through six quarts of *Turbo-Rocket* fluid in less than 200 miles of spirited driving!

Neither of the two cars would give genuine supercar acceleration. Not enough cubes or boost. But they were as fun to drive as anything Detroit offers today. Here are some standing-start times from my own road-test files:

	CORVAIR	**JETFIRE**
0–30 mph	3.6 seconds	3.5 seconds
0–60 mph	10.4 seconds	8.8 seconds
1/4-mile e.t.	18.1 sec @ 77 mph	16.9 sec @ 83 mph
Est. net HP	110	160

The Jetfire was a short-lived project, dropped in '63. But Chevrolet continued developing the Corvair turbo, eventually increasing displacement to 164 cubes. In '65–'66—the last years of production—the blown 'Vair was pumping out 180 HP, or about 1.1 HP per cubic inch! A *CAR and DRIVER* road test of a '65 Corsa showed that the turbo engine could propel the 2750-lb car to 60 mph in 10.2 seconds. 1/4-mile e.t. was 17.2 seconds at 78 mph. While this wasn't supercar acceleration, it certainly wasn't a bad showing for 164 cubic inches. Incidentally, instrumentation on the Corsas was respectable with a cylinder-head-temperature gage and a manifold-pressure boost gage.

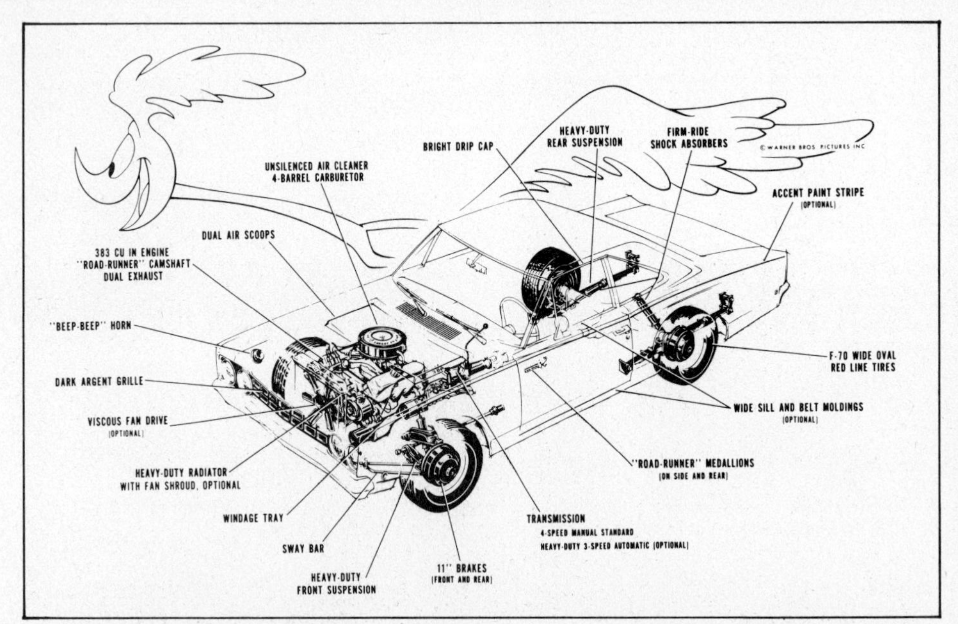

Plymouth Road Runner, introduced in 1968, was the most-successful budget supercar. Built from the intermediate B-Body coupe, it had a special high-performance 383 engine, heavy-duty suspension, and plenty of wild graphics. Photo courtesy of Chrysler Corp.

Coyote Duster cold-air system for the '69 Road Runner used a big rubber gasket to seal the open air cleaner and air housing around the 4-barrel carburetor.

Road Runner concept gave supercar guts without all the frills. Some would go as far to say that it was a taxicab with a big engine. It sold for $600—$1000 less than most supercars. Photo courtesy of Chrysler Corp.

Dodge called their B-body budget supercar the *Super Bee*. Same equipment as the Road Runner, but with bolder graphics. Didn't have the "meep-meep" horn nor did it sell as well as the Road Runner. Photo courtesy of Chrysler Corp.

Admittedly, the test car was an unusually good-running Sprint. Other magazine tests generally showed quarter-mile times around 17.2 seconds and 160—180 net HP.

Tempest Sprints were fun cars to drive—primarily because of the eager willingness of the engine to rev. I've seen 7000 rpm on brand-new models, before the valve springs lost some of their force. But actually, the engine's breathing wasn't efficient enough to make more power in these speed ranges. Best acceleration times were made by shifting at 5500—6000 rpm. It was a real problem to street racers who liked to make good noises as well as win races!

What killed the overhead-cam Tempest was its high manufacturing cost. The Sprint was only $300 less than a GTO, and it cost more than a similarly equipped Tempest with the 326-CID 4-barrel V8. And the V8 Tempest was just as quick. Besides, who wanted to be seen with a six?

Pontiac soldiered on with the OHC Six through '69, offering it in the Firebird as well. In '68, the cammer was stroked to 250 CID. The highest specific output came in that last year of production—a whopping 230 HP at 5400 rpm. Not too shabby for a six-banger.

1967 Buick California GS—This car was the idea of Buick's West Coast sales manager, Mickey Garrett, not something born in Buick's engineering department. Garrett thought there might be a substantial market for a stripped Skylark coupe with the 340-CID small-block V8—with enough special trim to *look* like a big-block GS. It was strictly a trim package, nothing more.

The 4-barrel 340 had no special performance parts—not even a chrome air cleaner. There wasn't anything underneath either: no police suspension, wide-rim wheels or premium tires. But the blacked-out grille, rally wheels and fender emblems were very much in the youth-market vein. And the base price of $3128 satisfied skeptical Buick dealers. There was considerable interest early in the model year. Unfortunately performance didn't match the image:

0—30 mph	3.7 sec
0—60 mph	9.3 sec
1/4-mile e.t.	16.7 sec @ 81 mph
Est. net HP	170

Dodge introduced the compact Dart Swinger 340 in 1969. It was about the same size and weight as the Barracuda, gave similar performance. But it didn't have the sleek styling image. This is a '70 model. Photo by David Gooley.

Dodge Super Bee body graphics consisted of a bumble-bee emblem and rear stripe. Photos by Ron Sessions.

The tests showed decent acceleration for a small-block engine, two-speed automatic transmission and 2.78:1 final drive. But the California GS didn't have enough gimmicks going for it. In those days a youth car needed a lot of charisma—or it had to be able to prove itself at the stoplight. The California GS had neither.

1968 Plymouth Road Runner, Dodge Super Bee—The 440-CID, intermediate-body Plymouth GTX and Dodge R/T that Chrysler introduced in 1967 were some of the smoothest, best-performing supercars of the era. But they were also rather expensive. Road Runners and Super Bees debuted a year later as a serious attempt to offer almost as much image and performance for a lot less money.

The Road Runner/Super Bee formula was simple: Use the intermediate B-bodies, but strip them of frills like bucket seats, center console, luxury upholstery roll-up quarter windows and so on. Use the 383-CID low-block version of Chrysler's B-block—but hop it up with the big-port heads, high-flow carburetion and long-duration camshaft developed the previous year for the 440 Magnum engine.

Chrysler didn't use all the 440 bottom-end goodies to allow 6500-rpm speeds. But the improved breathing justified a rating of 335 HP at 5200 rpm for the new high-performance 383—compared with 300—330 HP for standard 383 4-barrel engines. The difference in net power was a lot more than this. In fact, in terms of net horsepower *per pound of body weight,* the new budget supercars were almost equal to the luxury GTXs and R/Ts.

To add recognition to the performance of the new cars, Chrysler relied heavily on body graphics. Decals and stripes were cheap, and Chrysler sales people used them to their fullest on the Road Runner and Super Bee.

For Plymouth, the Road Runner theme was developed around the famous Warner Brothers cartoon character. Remember the wild desert-bird logo on the side of the Road Runner? Road Runner cartoons were very popular in theaters and on TV at the time. They even provided a special horn that came close to reproducing the "meep, meep" sound in the cartoons. Absolutely preposterous, but try to find one of those horns in a junkyard.

The cars actually caught on strongly in the youth/performance market in '68—'69. Plymouth sold 128,000 in those first two years before the novelty tapered off.

Dodge never did anywhere near as well with the Super Bee. Though mechanically identical to the Road Runner, the graphics didn't come off as well. And there just wasn't the background image. But Road Runner sales made up for it. Chrysler sales people considered their econo-supercar program a real scoop.

It wasn't all looks and image either. The cars performed. CAR LIFE tested a Road Runner and Super Bee that were identical except for final-drive ratio. Both had automatics. It's interesting to compare acceleration figures:

	3.23:1 rear	**3.91:1 rear**
0—30 mph	3.1 sec	2.8 sec
0—60 mph	7.3 sec	6.6 sec
1/4-mile e.t.	15.3 @ 91 mph	14.7 @ 95 mph
Est. net HP	280	280

All this for a price about $500 less than a GTX or R/T. Supercar bargain!

1969 Dodge Dart Swinger—The Swinger—and its luxury-version GTS sibling—was Dodge's answer to the Barracuda 340 S. It used the standard Dart A-series hardtop body with the new high-performance 340-V8 engine. A minimum of luxury features held the package price under

One of the sweetest street engines of the supercar era was the 340-CID high-performance version of the Chrysler small-block V8. It combined power and quick response with good fuel economy. Photo courtesy of Chrysler Corp.

Big 4-in. flexible air tubes and dual-snorkel air cleaner gave an impressive underhood look to the 350-CID W31 Olds Cutlass. Air scoops of these '68—'69 models were under the bumper.

Olds first offered the 350-CID W31 high-performance engine in the '69 Cutlass. They called it the *RamRod* package. It would wind to 6600 rpm, and accelerate nearly as fast as the 400-CID 4-4-2. Photo by Leavenworth.

$2900, and the weight to 100-lb less than the 'Cuda. Yet the package pushed all the youth/performance *hot buttons:* Swinger-only red paint, fake air scoops on the hood, wrap-around *Bumble-Bee* stripes at the rear, and special low-restriction dual mufflers with *engineered Swinger sound!* It had good image per dollar.

Chrysler designed a special high-riser intake manifold for the 340, flanged for a 650-cfm Carter AVS 4-barrel. It was an excellent induction system. Photo by Mike Crawford.

Acceleration was a tad better than the 'Cuda 340 S because of the reduced weight. CAR LIFE recorded the following with Torqueflite and 3.23:1 final drive:

0—30 mph	3.0 sec
0—60 mph	6.9 sec
1/4-mile e.t.	14.8 sec @ 96 mph
Est. net HP	290

The Swinger's automatic transmission was calibrated for full-throttle upshifts at around 4800 rpm—well below the peak of the power curve. Best acceleration times were achieved with manual shifting at 5400 rpm. Chrysler engineers didn't always calibrate their high-performance cars to upshift at the top of the curve. It depended on whether the transmission had heavy-duty internal mods. This one was a medium-duty version of the A-727 Torqueflite.

Another interesting point made by the testers: This was one of the first performance cars to use the new bias-belted tires from Akron—specifically, Goodyear Polyglas 70-series. These tires not only improved handling and steering response, but dry-road traction seemed better. The testers took time to repeat the drag-strip tests with conventional bias-ply tires of the same size. They were surprised to note that the car was slower by 0.3 seconds. Traction was reduced so much with the bias tires that the driver had to launch at idle speed, rather than torquing up the rear suspension with the belted tires. And in fast driving, the Swinger was quicker than the Barracuda with 70-series bias tires. Think how it would have performed with today's radial tires!

1969 Olds Cutlass W31—This was a *real* budget supercar—in that the only special body trim and model identification was a couple of *W31* emblems on the fenders. Actually the W31 designation wasn't a separate model, but an optional engine/suspension package for any Cutlass coupe. And at an option price of only $205, it was one of the best bargains of the supercar era.

The engine was essentially a low-block 350-CID version of the W30 *Force-Air* package developed for the 400-CID 4-4-2 in 1968. This included under-bumper air scoops feeding outside air to the carb through 4-in. flexible hoses. It was probably worth another 10—15 HP above 80 mph.

Olds engineers wanted to use the big-port 400—455 heads on the 350 block for the W31 combination. But dyno

The only identification on the '68 W31 Cutlasses were tiny *RamRod* decals on the front fenders. It was a rush job to get the package out on the streets, to compete with the Road Runner.

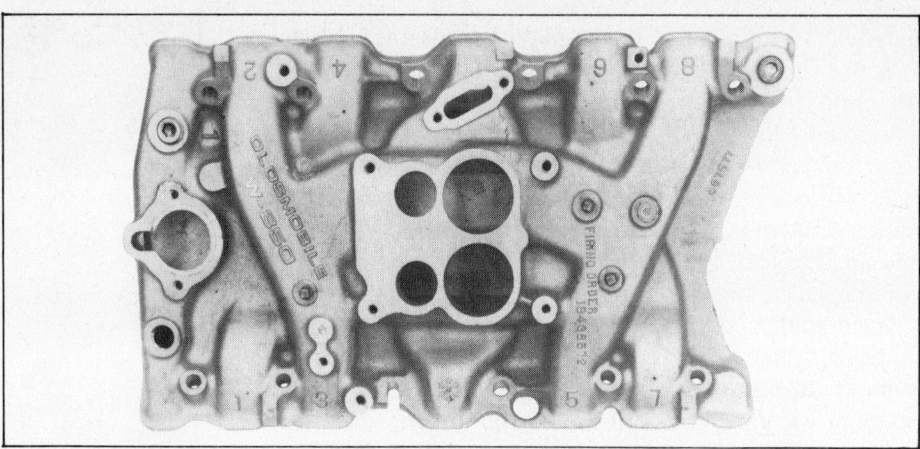

Many car fans don't know that the 1970 W31 Cutlass engines had aluminum intake manifolds. So did the 455-CID W30 jobs. Olds also offered an aluminum center section for the rear axle. Weight was beginning to get some attention at that time. Photo by Leavenworth.

tests showed these ports were *too big* for smooth street running with high-performance cam timing. They ended up using the small-port 350 heads, but with 1/8-in.-larger intake valves. Then the engineers slipped in a long-duration 308° cam with 0.47-in. lift to boost power above 5000 rpm. Meanwhile the lightweight valve train allowed the engine to turn an unprecedented 6600 rpm with hydraulic lifters and medium-rate valve springs. A side benefit of the medium-rate springs was that they didn't cause premature cam-lobe wear.

It was one of the more-brilliant engineering combinations to come out of the supercar period. The W31 had a *fat* top end with smooth, flexible street cruising. Certainly a much-better combination than, say, the Boss 302 Mustang—which had too much intake-port area.

This uncanny 6600-rpm revving ability and the unusual top-end breathing with the long-duration cam and efficient cold-air feed gave W31s important performance advantages over other hydraulic-cam supercar engines. For one thing, being able to rev well over 6000 rpm reduced the acceleration loss from the wide gear ratios in the Turbo Hydra-matic 350 transmission. The W31 didn't need a close-ratio four-speed like many other "peaky" engines. Olds engineers used a high-stall torque converter for getting off the line. But from 3000 rpm up it was Rev City.

How many other small-block engines could produce more than 300 net horsepower as installed in the car? The result was acceleration with an intermediate body as good as other engines pulling much-lighter compact cars. CAR LIFE testers were impressed with these figures, using a Turbo Hydra-matic and 3.91:1 final drive:

0—30 mph	3.0 sec
0—60 mph	6.6 sec
1/4-mile e.t.	14.9 sec @ 96 mph
Est. net HP	310

One interesting result of the long cam duration with increased overlap is that there wasn't enough intake-manifold vacuum to operate a power-brake booster. This unexpected development almost killed the W31 program.

AMC SC/Rambler of '69 was perhaps the most potent of the junior supercars. Each Hurst-modified Rambler Rogue had a high-performance 390-CID V8, traction bars, ram-air hood and much more. Cars were capable of low-14-second quarter-mile times with trap speeds near 100 mph—off the showroom floor. Photo courtesy of Hurst Performance, Inc.

Olds didn't want to spend money on a separate engine-driven vacuum pump. They finally bit the bullet and released the car without a power-brake option. Olds tried to compensate by using soft brake linings requiring a mere 90-lb pedal force for a high-G stop. But the price they paid was a higher fade rate under hard use and frequent lining changes. Owners complained—until they punched the throttle and felt that crazy cam haul on the top end. Interesting car.

1969 AMC SC/Rambler—They say this car was really the idea of George Hurst, the man behind Hurst Performance, Inc. Hurst not only sold American Motors officials on the package, but he offered to build the cars from a standard AMC production combination in his spacious Detroit-area prototype shop. The deal was much like custom fabricators who build convertibles by chopping the tops off production coupes.

AMC liked the idea. It enabled them to get into the junior-supercar market without all the tooling cost and fuss. The AMC factory produced a batch of Rogue coupes with big-displacement 390-CID 4-barrel V8s. The cars also had factory options like heavy-duty suspension, Goodyear Polyglas 70-series red-line tires on 14 X 6-in. mag-style Magnum 500 wheels, Warner T-10 four-speed transmission, 3.54:1 limited-slip rear end, and so on. Then the cars were shipped to the Hurst shop. Hurst gave the cars a wild red, white and blue paint job, Hurst T-handle shifter, fender emblems, rear traction bars from the AMX chassis, and an excellent hood scoop with sealed air cleaner to feed cold ram air to the carb.

Incidentally, the Hurst SC/Rambler had one of the better air-feed systems of the time. A vacuum-operated valve in the scoop opened at wide-open throttle. The scoop opening was far enough above the hood surface to avoid boundary-layer effects. Plus it was impressive looking!

It was inevitable that the SC/Rambler would be a star performer. Curb weight was only 3160 lb. The big-displacement V8 engine put 61% of this weight on the front wheels, but somehow the traction bars and 70-series belted tires gave decent off-the-line traction. And the efficient cold-air system and dual low-restriction exhaust assured a net power rating very close to the advertised 315 HP. CAR LIFE recorded acceleration figures for the SC/Rambler that were the best of all the budget supercars:

0—30 mph	2.4 sec
0—60 mph	6.3 sec
1/4-mile e.t.	14.2 sec @ 100 mph
Est. net HP	300

This performance was all the more remarkable because the standard lifters and valve springs in the engine limited revs to about 5000 rpm. You couldn't wind up over the power peak in the gears. But the fat mid-range torque with those 390 cubes seemed to compensate. You could get neck-snapping acceleration at any speed with the four-speed.

Terrific little car, the SC/Rambler. And believe it or not, it listed for only $2998!

1970 Plymouth Duster 340—When Chrysler introduced their new E-body ponycars in 1970, they wisely retained some A-body compact coupes for the budget-supercar market. Plymouth's version was the Duster. The concept behind the Duster was not unlike that of the earlier Barracuda: a sporty model based on a bread-and-butter coupe. The difference was that the Duster line started around $2500. That price got you a plain-Jane six-cylinder *strippo* with bench seats and three-speed column shift. Everything above that cost extra.

The high-performance 340 came as part of a package that included dual exhaust, heavy-duty suspension, wide-rim wheels, 70-series belted tires, front disc brakes, three-speed floor shifter and a rally instrument cluster. At $400, it was a bargain. You could have a strong street racer for less than three bills.

As expected, performance was similar to the 340-CID Barracudas and Dart Swingers of the late '60s. CAR LIFE recorded the following with Torqueflite transmission and 3.23:1 final drive:

Rallye 350 rear spoiler was said to help stability at high speeds—130 mph with the W31 engine! Photo courtesy of Oldsmobile.

0—30 mph	2.6 sec
0—60 mph	6.2 sec
1/4-mile e.t.	14.7 sec @ 94 mph
Est. net HP	270

The CL testers' biggest criticism was poor handling, considering the heavy-duty suspension. Apparently, spring and shock stiffness were compromised a lot to get a less-harsh ride because the Duster was aimed at a broader market segment than the earlier 'Cudas and Swingers. However, the front disc brakes that came standard with the 340 package gave outstanding stopping performance.

Plymouth dealers sold more than 25,000 Duster 340s that first year. Some dealers felt the car stole potential sales from the higher-profit E-body 'Cudas. There was much controversy about the car in Chrysler's front office. But they continued producing the line through late 1973—when a less-potent 360 V8 was substituted. The 340 Dusters probably did more for the MoPar street-performance image than the big-block 'Cudas.

1970 Olds Cutlass Rallye 350—It's hard to trace the real roots of Olds' wild Rallye 350 package of the '70—'71 period. Some say it was an attempt to outshine the super-bold paint and graphics of Pontiac's '69 GTO Judge. Some say it was a last-ditch attempt to attract a hard core of youth/performance buyers jaded by all the styling cliches, decals and gimmicks of the '60s. Anyway, the Rallye 350 had to be a high-water mark in supercar styling.

Picture it: A standard Cutlass coupe with a bright Sebring-yellow paint, with urethane-coated front and rear bumpers and rally-wheel inners to match, and bold black-and-orange striping for accent. But there's more. A neat wing over the rear deck provided downforce. In front, a new fiberglass hood with two huge air scoops extended to the leading edge. Rallye 350s were different-looking cars. Many thought they were beautiful. Take a good look at the pictures before you condemn Olds stylists for overdoing it. For sure, you couldn't miss these cars on the street!

Plymouth restyled the Valiant two-door coupe body for 1970, and called it the *Duster.* With optional high-performance 340 V8, '71 Duster could out-accelerate heavier 400-CID 'Cudas. Photo courtesy of Chrysler Corp.

There was more than crazy paint and spoilers. Olds engineers tried a new performance tack with the Rallye 350. You recall they used bigger valves and a very radical camshaft to get power at high revs on the earlier Cutlass W31 package. But the W31 paid for it in reduced low-end torque, rough idle, some fuel-economy loss and *no power brakes!*

For the Rallye, Olds plugged in the standard 350 4-barrel engine—standard valves and cam—but used the W31 cold-air system and low-restriction dual exhaust to get as much *free* horsepower as possible. And it worked. The Rallye came within 10—20 HP of the W31 with considerably better street flexibility and fuel economy—plus a power-brake option. There was more difference with open exhaust, where the W31 may have been 40- or 50-HP stronger.

Oldsmobile's cold-air system was changed radically in 1970. The late-'60s system used under-bumper air scoops and 4-in.-OD flexible hoses leading up to the air cleaner. It was efficient, but it was a headache to install on the assembly line. The new system used a styled fiberglass hood with large twin air scoops. The scoops fed to a rubber-sealed air-cleaner housing that had a vacuum-operated trap door to let in cold air only at wide-open throttle. Under cruising conditions, the car took in warm underhood air. This helped driveability in cool weather. Whether the late system was as efficient from a standpoint of ram pressure at high speeds, I don't know.

Anyway, the Rallye 350's acceleration was quite commendable—considering the standard engine and 3700-lb curb weight. CAR LIFE testers obtained these figures with the Muncie four-speed and 3.42:1 final drive:

0—30 mph	2.8 sec
0—60 mph	7.0 sec
1/4-mile e.t.	15.2 sec @ 94 mph
Est. net HP	300

The high net power was undoubtedly due to the cold-air feed, efficient dual exhaust, and the fact that CL testers were able to use shift points up to 5300 rpm with the standard valve springs. Remember, the W31s would go to 6600 rpm with medium-rate valve springs. It was an excellent showing for a basically standard bread-and-butter engine. Testers also spoke well of the handling with the rear anti-roll bar used on all Olds supercars. The new front disc brakes—power-boosted this time—provided good stopping power from 80 mph.

The Rallye 350 was a lot of car for an option-package price under $500.

WERE THEY SUCCESSFUL?

At the beginning of the chapter, I stated a prime rule of marketing; *Reducing the unit price of an item broadens its potential market.* According to that theory, the junior supercars described here should have easily outsold the conventional GTO-type supercars.

They didn't. They didn't come anywhere near it! With few exceptions, company sales executives were disappointed with their budget-supercar programs. Profits didn't cover tooling, merchandising and distributing expenses in most cases. I don't have any hard-and-fast dollar figures to prove it, but the Plymouth Road Runner is the only specific model listed here I'd consider a genuine Detroit-style money-maker. Marginal possibilities would be the Duster/Swinger 340s.

Of course, some of the cars listed were option packages on a base model. Sales figures are rarely published on option packages. But volumes were never impressive by Detroit standards. As an example, Oldsmobile's W31 package for the Cutlass sold about 3000 units from late '68 through '70. They sold more than twice this many W30 packages for the 4-4-2 during this period—and 4-4-2 model sales were running about 30,000 units a year at the time.

The implication is obvious. The kids weren't trying to pinch pennies with their street machinery in those days. They wanted bucket seats and floor consoles and tachometers and fancy wheels and body trim—and they were apparently willing to sell their souls to pay the price. That $500 to $800 or so difference between the budget cars and the real supercars was apparently *not* a key sales factor. The kids still bought cars that had image and performance. No budget supercar had a better combination of both than the Road Runner. And look how it sold.

But the fact remains that the junior-supercar market gave us some interesting, exciting cars to remember. Because most of them were compacts with small-block engines, their general handling and road agility were better than the intermediates with big-blocks. The *juniors* were fun to drive. And acceleration performance was very close to the big-block supercars—e.t.s in the low 15s and high 14s were the norm. Wouldn't it be nice if one could buy that kind of four-wheel performance for $3000 today?

Corvette Tri-Carb system was immediately indentifiable by its unique triangular air cleaner. Photo by Ron Sessions.

For most of its existence, the Corvette has been the only volume-production U.S. sports car. From the very start, it offered luxury uncommon in most imported sports cars. Photos by Ron Sessions.

American Sports Cars

A breed apart

ANOTHER BREED

No book on American supercars would be complete without looking at the two-seat, open sports cars that have maintained a small niche in the U.S. market. Perhaps it's not fair to compare them directly with full-bodied muscle cars like the GTO. Two-seat sports cars have traditionally been smaller and lighter, with better weight distribution, more-sophisticated suspensions and more-costly equipment. They *should* have been quicker and faster.

But these obvious facts were no consolation to the proud owner of a Street Hemi who had just been dusted by a 427 tri-carb Corvette. He had to live with these nasty little 'Vettes back in the '60s.

In those days there was constant war between the devotees of conventional muscle cars and the *sports-car set.* They were two distinct camps. No quarter was asked, and none given. If you couldn't handle a certain Corvette or Cobra with your Six-Pack Road Runner or Boss 302, you started planning modifications. More carburetion, more cam, tubular exhaust headers, fatter tires, more gear. No muscle-car fan worth his salt would concede defeat just because the two-seaters weighed 500—1000-lb less. Hot-rod shops were ready to sell bolt-on goodies that they claimed would make up the difference. You went down with your foot on the floor and engine screaming!

Above: Original six-cylinder Corvette of 1953 offered mild performance and handling. It was mostly an image car for the GM corporation. Sports-car purists laughed at it. Photo by David Gooley.

CHEVROLET RAISES THE STAKES

The idea of a U.S.-built, open two-seat sport runabout was not new when Chevrolet introduced the Corvette in 1953. We learned in Chapter 1 that such cars have had a lot of buyer appeal from the earliest days of the automobile.

But it was only appeal. Most of these U.S.-built two seaters of decades past were too expensive for any but the rich. Attempts to market small, low-priced sports cars after World War II—like the Crosley Hotshot, Kaiser-Darrin or Nash-Healey—were doomed to failure. The cars simply didn't offer superior performance. The concept was lost somewhere in the attempt to cut costs.

Chevrolet's Corvette would have bombed too if GM hadn't put the new small-block V8 engine in it as soon as it became available in 1955. The original 'Vette used the old "stovebolt" six-cylinder with three side-draft Carter carburetors. They backed it up with the lazy two-speed Powerglide automatic. Acceleration of the '53—'54 model was no better than many standard sedans of the day. It was a fun car to drive and flashy to be seen in—nothing more.

But auto historians generally agree that superior performance has been a major part of the Corvette mystique down through the years. That performance mystique was assured when the 265-CID V8 replaced the old stovebolt. And Zora Arkus-Duntov—"father of the Corvette"—guarded that performance edge jealously for the next 20 years. He was appointed chief engineer for the Corvette program in the mid-'50s. His stamp was on just about everything about the car until his retirement in the

Zora Arkus-Duntov, shepherd of Corvette development for 20 years, drove this prototype '56 Corvette to an average speed of more than 150 mph on Daytona Beach. He was always alert to publicity stunts that would draw attention to his "baby." Photo by Wally Chandler.

Engine in Duntov's 150-mph Corvette was a 265-CID dual-quad '56 model with an experimental "Duntov" cam that was to be standard equipment later. It developed 240 HP at 5800 rpm. Photo by Wally Chandler.

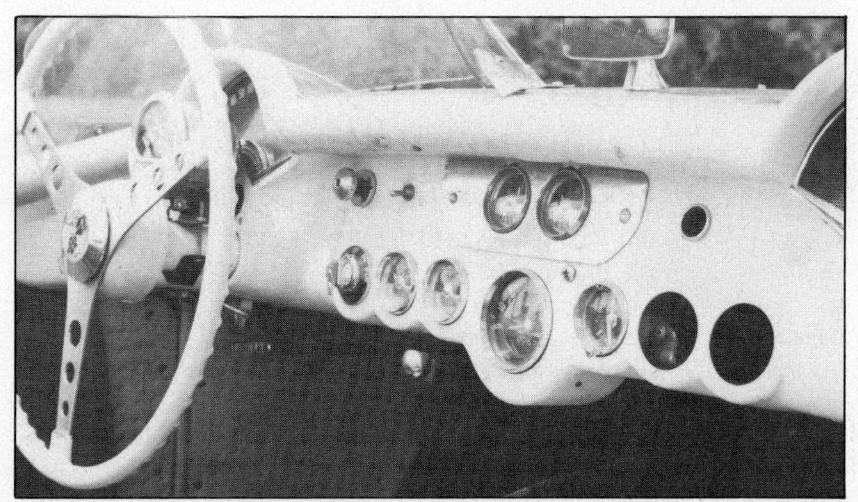

Duntov cut a speed record up Pikes Peak in late '55 with an experimental close-ratio three-speed floor-shift transmission. It showed the need for the new four-speed that was introduced in '57. Photo by Wally Chandler.

Corvette engineers adapted a Bendix ceramic/metallic brake lining and finned drums for road racing in 1956. These were offered as a regular-production option in the late '50s. Photo by Wally Chandler.

mid-'70s. The Corvette was Duntov's official entry in the '60s supercar sweepstakes.

I think it's appropriate to say a few words here about Duntov's philosophy of sports-car design. After all, his Corvette pretty much set the U.S. standard for two-seater performance during the period we're studying. For one thing, Duntov was always concerned with *total* performance: braking and cornering as well as straight-line speed and acceleration. This is why much of the Corvette's development budget was spent on brakes, suspension, tires, transmissions—not only on horsepower.

Weight was another bug to Duntov. He detested every pound that was added to the Corvette from year to year—especially when it was to add something as unimportant as thicker carpeting or an FM-stereo radio. For years, Duntov fought putting a big-block V8 in the 'Vette. It wasn't until the Mark IV big-block was available in 1965 that he finally gave in. And that was only because the new '63 Sting Ray chassis had independent rear suspension and considerable engine setback so 50/50 weight distribution could be maintained with the heavier engine.

Due to this Duntov design philosophy, the Corvette would maintain all-around performance potential competitive with some of the most-expensive and exotic sports cars in the world. And it's not surprising that his Corvettes were tough for full-bodied muscle cars to beat, even in straight-line acceleration. Duntov tried to keep the Corvette out front in *every* aspect of performance.

FORD ANSWERS THE CHALLENGE

No sooner had the first Corvette hit the streets in 1953 than Ford Motor Co. officials huddled to see what they could come up with to meet the new challenge, and possibly make some money in a brand-new market segment.

Ford's Thunderbird of 1955 was an answer to the Corvette. But it sold so much better than the 'Vette that Chevy officials considered dropping their fiberglass two-seater. Photo by David Gooley.

Whereas the Corvette made pretensions at being a true sports car, the two-seat Thunderbird clearly took its styling cues from big cars. Photo by Ron Sessions.

Their Thunderbird two-seater of the '55—'57 period was the result. It was Ford's concept of what an American sports car should be.

Obviously that concept was a lot different than Duntov's. The new Thunderbird was a lot more plush and fancier than the Corvette of that period, weighing 300—400-lb more. One unique Ford idea was to offer an optional hard-top roof section that could be fastened on to form a snug coupe for winter, and then be quickly removed to make a convertible in summer. It was a popular option. In fact, the Thunderbird had more market appeal than the Corvette. It outsold the 'Vette 4-to-1 in the '56—'57 period. If Ford officals had any apprehensions about the importance of performance superiority, they apparently felt their new *Y-block* V8 would keep them in the race.

It's interesting to see how Corvette and Thunderbird performance compared during this time. Below are some results of ROAD & TRACK Magazine tests—plus one test from my own files.

Obviously Chevrolet did a better job developing Corvette performance than Ford did with the Thunderbird in those early years. But in defense of Ford, they were shooting at a different market with the 'Bird than Chevy was with the 'Vette. Ford was after a luxury/prestige segment of the two-seat market: Duntov was after the genuine sports-car fan. Ford engineers made no attempt to control

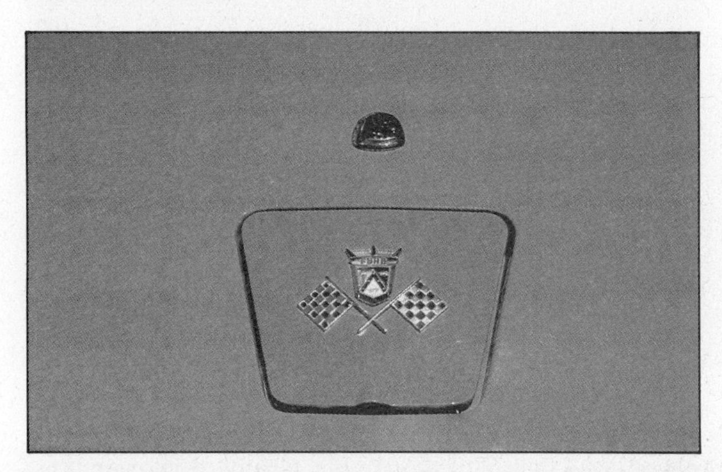

How to tell a '55 T-Bird from a '56. The '55 had crossed flags under the Ford logo and dual "gunsight" exhausts through the bumper. Photos by Ron Sessions.

Car	Transmission	Axle ratio	Adv. HP	0—30 mph	0—60 mph	1/4-mile e.t. @ mph	Est. Net HP
'55 Corvette 265	A3	3.55:1	195	3.2 sec	8.7 sec	16.5 @ 84	170
'55 T-Bird 292	A3	3.31:1	198	3.7	9.5	17.1 @ 78	150
'56 Corvette 2x4 Bbl	M3	3.55:1	225	2.7	7.3	15.8 @ 89	210
'56 T-Bird 312	A3	3.31:1	225	3.3	9.3	17.0 @ 82	190
'57 Corvette 283 FI	M4	4.11:1	283	2.5	5.7	14.3 @ 95	240
From Personal Tests:							
'57 T-Bird 312 Supercharged	M3	3.89:1	300	3.0	7.1	15.6 @ 90	250

1956-'57 Thunderbird 312-CID 4-barrel engine was dressed up with beautiful valve covers and air cleaner. But it couldn't match the power of the 283 Corvette and had to move several-hundred pounds more weight to boot!

Ford tried to compete with Corvette performance in the '55—'57 period with their new Thunderbird two-seater. They did fairly well at the Daytona Beach speed runs with a 312-CID engine and Paxton supercharger. Photo by Wally Chandler.

the T-Bird's weight. In 1955, the 'Bird was 350-lb heavier than a V8 Corvette. By '57, it was 600-lb heavier! There was no way its acceleration could keep up.

The 312-CID supercharged T-Bird engine of 1957 was only a token gesture at performance. Perhaps it would be more fair to say the blown engine was never developed to its full potential. The infamous AMA anti-racing resolution came down only a few months after the engine was introduced. And that put a stop to this type of development at Ford Engineering. Only a few T-Birds and Ford sedans were sold with the supercharged engine in the early months of 1957.

Then in 1958, Ford introduced an entirely new four-seat Thunderbird personal-luxury car—and got out of the sports-car business altogether.

CORVETTE FIGHTS THE WEIGHT BATTLE

The '57 283-CID fuel-injected Corvette was a performance milestone from several standpoints. More displacement, improved cylinder heads and the high-performance *Duntov* camshaft produced another 30—40 HP and better mid-range torque. The constant-flow fuel system didn't add a lot of top-end power, but instant throttle response and positive fuel feed in the corners helped road-circuit performance considerably.

The new Warner T-10 close-ratio four-speed aided acceleration by allowing engine revs to be kept closer to the power peak when shifting gears. When combined with optional 4.11:1 final drive, you had a beautiful ratio spread for standing-start acceleration. And perhaps the most-important point in the '57 FI package: Curb weight with full equipment was still under 3000 lb.

As the road tests show, with the '57 FI setup, Corvette performance reached a peak that remained unsurpassed for several years.

The bug was weight. What happened was that the inev-

Fuel-injected Corvette engine introduced in 1957 was one of the first American production engines to claim 1-HP per cubic inch—283 HP at 6200 rpm, a real breakthrough in performance. Photo courtesy of Chevrolet.

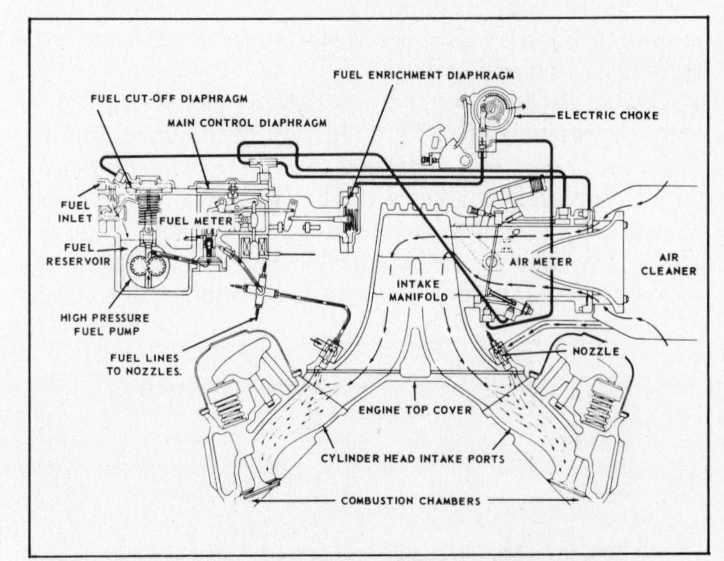

Rochester fuel-injection system for 1957 Corvettes used a venturi-type air meter to give a vacuum signal proportional to air mass. A diaphragm controlled the air/fuel ratio. The air manifold had long ram passages. Drawing courtesy of Chevrolet.

Duntov made big changes to the Corvette in the late '50s. Styling changes were evolutionary, but performance was improved radically. Fuel-injected models gave 14-second quarter-mile e.t.s with only 283 cubes! Photo by David Gooley.

1957 Corvette is considered by many enthusiasts to be the cleanest design of the live-rear-axle—'53—'62—models. Photo courtesy of Cars & Parts Magazine.

Fuel-injection script on fender scallop was the mark of the fastest small-block Corvette. Photo courtesy of Cars & Parts Magazine.

Last of the live-rear-axle Corvettes were popular drag-racing machines. Small-block 327-CID V8 was available in '62. Photo by Tom Monroe.

Intake ports were enlarged considerably on experimental Corvette aluminum heads in 1960. This port design was adopted as standard equipment for most later 327 and 350 engines—but in cast iron.

itable upgrading of the car to appeal to a broader market—additional equipment, fancier upholstery, more body trim, stronger chassis components to handle the added weight—increased the weight as fast as the engineers could improve the power and torque output of the small-block engine. It was a vicious cycle. This is shown dramatically by comparing ROAD & TRACK tests on several models between '57 and '65. Weight increased as fast as power did, *so performance stood still.* In each case, I'm comparing four-speed cars equipped with the strongest FI engine available.

Below, you can see that the increases of 60 HP and 44 cubic inches (283 to 327 CID) were just about offset by the 250—300-lb increase in weight. It was a tough morsel for Duntov to swallow. But *weight escalation* is just about inevitable when you tailor a car to appeal to a broader market. In fact, it's remarkable that the weight increase of the new independent rear suspension and Sting Ray coupe body in

Year	Changes	Curb weight	Adv. HP	0—30 mph	0—60 mph	1/4-mile e.t. @ mph	Est. Net HP
1957	Baseline FI	2880	283	2.5 sec	5.7 sec	14.3 @ 95	240
1961	Bigger ports & valves	3080	315	3.1	6.6	14.2 @ 98	280
1965	Sting Ray coupe body, more displacement, more cam	3150	375	2.9	6.3	14.4 @ 99	300

Last fuel-injected Corvette for nearly two decades appeared in '65. Displacing 327 cubes, it put out an advertised 375 HP. It was losing popularity due to the introduction of the big-block 396 V8. Photo by Tom Monroe.

Introduction of the Sting Ray body in '63 put Corvette at the cutting edge of styling. Photos by Ron Sessions.

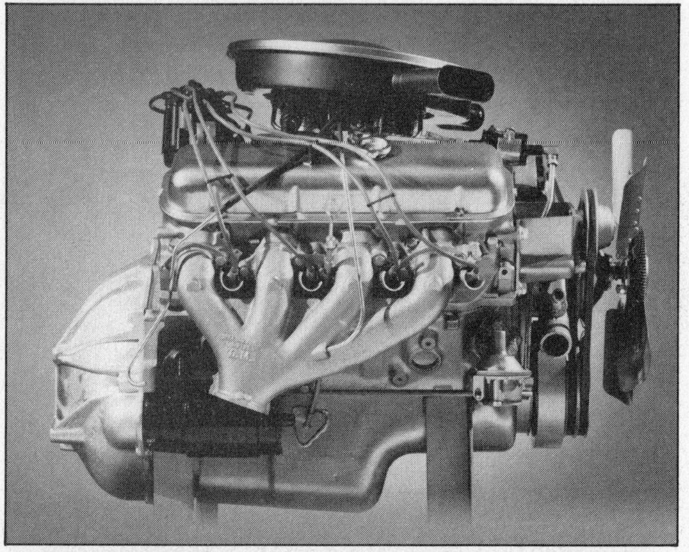

Big-block Mark IV high-performance engine was first offered in 1965 Corvettes. This first version displaced 396 cubic inches, rated at 425 HP. Those beautiful exhaust manifolds were only available on Corvettes. Photo courtesy of Chevrolet.

In 1967, the Corvette became the first U.S. production car to offer four-wheel disc brakes as standard equipment. Photo by David Gooley.

1963 was less than 100 lb. It speaks well for the weight-consciousness of the 'Vette engineers.

At this point, there was really only one answer for maintaining Corvette performance.

A BIG-BLOCK FOR THE 'VETTE

Chevrolet introduced their first big-block V8 in 1958—an unusual design with wedge-shaped combustion chambers *in the block* and staggered valve positions in the head. It's known as the *W-block.* The high-performance 409-CID version was quite a stormer in super-stock ranks in the early '60s, as discussed in Chapter 3. But Duntov was utterly opposed to using this engine in the Corvette. He said it was too heavy for the power it produced. It would upset the car's good weight distribution. He fought tooth and nail against some strong feelings in the Chevy organization to keep the *W* engine out of the 'Vette. He succeeded.

But the introduction of the all-new Sting Ray model in 1963 changed the picture. The 100-or-so-pounds of additional weight required more cubes to maintain or improve performance. There was plenty of room for a big-block in the engine bay. And most important, the substantial engine setback and heavier independent rear suspension would permit another 150 lb on the front wheels and still have nearly 50/50 weight distribution.

Duntov still wouldn't have accepted the W engine. But introduction of the new *Mark IV* big-block in 1965 finally gave him the kind of power-to-engine-weight ratio he could accept for his beloved Corvette.

Not that the new Mark IV was a lightweight. The Mark IV weighed around 670 lb—almost as much as the W—but it had a lot more power and torque potential. The unique "porcupine" valve arrangement and large ports gave outstanding breathing. When the Chevy front office approved a fat budget for further performance development, the stage was set. From that time on, only conservative Corvette fans chose the small-block engine.

In the Corvette, the 425 HP, 396-CID big-block gave better acceleration than a $25,000 Ferrari. Side exhaust pipes were an option. Photo courtesy of Chevrolet.

When the big-block Mark IV V8 was added to the Corvette in '65, drag racers embraced the package. Interestingly enough, where allowed by class rules, the car's IRS was replaced by a live rear axle. It didn't have the geometry or durability for drag racing.

That first 396-CID Mark IV high-performance engine was a tremendous performance breakthrough for the 'Vette. Special engine goodies included big-port cylinder heads, a beautiful high-riser aluminum intake manifold, 780-cfm Holley carb, forged 11:1 pistons, long-duration, solid-lifter cam, four-bolt main-bearing caps, heavy-duty rods, and so on.

Also, much attention was paid to induction and exhaust systems. This was the first engine to use the big 14 X 3 in. AC air filter. And the exhaust side had big, special cast-iron headers feeding into large-diameter headpipes and dual low-restriction mufflers. It was one of the best factory exhaust systems around.

Acceleration figures from ROAD & TRACK tests on a four-speed with 3.70:1 final drive confirmed the newfound power:

0—30 mph	3.1 sec
0—60 mph	5.7 sec
1/4-mile e.t.	14.1 sec @ 103 mph
Est. net HP	350

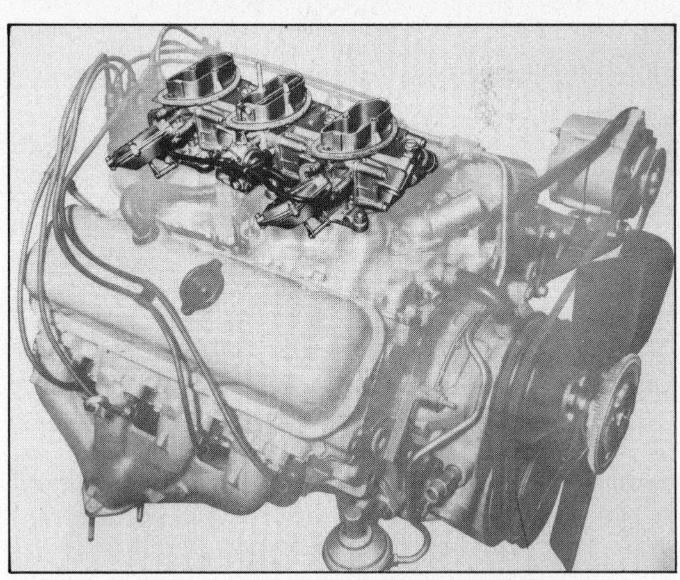

An interesting 6-barrel carburetion system was offered on 427 Corvette engines in 1967. It had a total airflow capacity of 950 cfm, added about 10 HP over the 780-cfm 4-barrel. Photo courtesy of Chevrolet.

Practically overnight, the Corvette quarter-mile performance bogie dropped to the very low 14-second bracket, with trap speeds consistently over 100 mph. Finally, here was a combination to match or better the performance of the '57 283-CID FI car. The extra 50 HP and gobs more torque from the Mark IV big-block did the trick—despite a 150-lb weight gain over the previous year.

This wasn't the end of Corvette performance development in the '60s. In 1966, the Mark IV block was bored out to 427 cubic inches. Chevy kept the same advertised power rating of 425 HP for the 427 4-barrel combination—otherwise identical to the '65 396 engine—but the extra cubes produced a little more mid-range torque. This 427 was a nice street engine.

The following year came the fabled 427 *Tri-Carb* engine. This carburetion system was described briefly in Chapter 4, as it was also a feature of the Dodge/Plymouth 440 Six-Pack engines. But the idea originally came from Corvette engineers, working with the Holley-carburetor people. Essentially what GM wanted was carburetion with more *image* than a single 4-barrel. They also wanted more flow capacity *and* equal smoothness and flexibility under cruising conditions.

What came out was a setup with three 2-barrel Holleys on a big-passage, two-plane aluminum manifold, with a huge, triangular foam-type air filter spanning all three carbs. The throttle linkage worked only the center carb. All idling and cruising was done on that one carb. The two end carbs were opened automatically by vacuum diaphragms that were actuated by airflow through the *center* carb. At

Corvette underwent a major restyle in '68. Basic '68 styling theme was retained for the next 15 years, with relatively minor changes. And the car was just as popular in the early '80s as ever. This is the '69 variation. Photo by David Gooley.

1967 427 Corvette remains the quickest Corvette ever in production. Photo by Ron Sessions.

wide-open throttle, they would open gradually between about 2000 and 4000 rpm. With all six barrels open, the three carburetors combined to give 950-cfm of airflow—about 70-cfm more than the single Holley 4-barrel on the '66 engine. This, plus the improved mixture distribution gave 10—20-more net horsepower above 5000 rpm.

It was an expensive carburetion system that required a lot of maintenance. But no one questioned its overall performance potential. And it had image in spades. If you drove up in a 427 'Vette, the first thing the street freaks looked for was those three deuces under the big triangular filter!

CAR LIFE Magazine ran a series of road tests in 1969 on all the popular Corvette engine and drive-train combinations. I'll refer to these to give you some idea of how Corvette performance developed in the late '60s. All the tests cars were '69 models, and weighed within a few pounds of each other. Here is a quick rundown on the four engine combinations:

Base engine (ZQ3): Standard 350-CID 4-barrel combination with standard heads, 10.3:1 compression, Quadrajet carburetion (750 cfm), standard low-lift cam and standard exhaust manifolds with dual exhaust. Rated 300 HP at 4800 rpm.

Option L46: Successor to the famous 350-HP 327-Corvette engine, but displacing 350 cubes, and with a mild hydraulic high-lift cam. It used the big-valve FI heads, 11:1 forged pistons, Quadrajet carburetion, forged crank, four-bolt main-bearing caps and other goodies. Capable of 6000—6200 rpm with fresh valve springs, it was advertised at 350 HP at 5600 rpm.

Option L36: *Civilized* version of the 427 high-performance combination, the L36 was essentially a standard Mark IV setup with standard small-port heads and 10.3:1 compression. It also had a Quadrajet carb on the cast-iron low-riser manifold, mild hydraulic high-lift cam and standard exhaust manifolds. The engine delivered considerably more power in the Corvette than in a passenger car because of the excellent dual exhaust system and the big open-element AC air cleaner. A closed snorkel-type cleaner was used with the L36 engine in passenger cars. Maximum rev capability: about 5600 rpm. Rated at 390 HP at 5400 rpm.

Option L71: This was the mighty 427 Tri-Carb engine, rated at 435 HP at 5800 rpm. Actually, the L71 used all the same components as the original high-performance 396 of 1965, except for the additional carburetion: The extra cubes and carburetion netted about another 50 HP. Maximum revs, still 6500 rpm. It was a bargain at $437 more than the base 350.

Here are the acceleration figures from the CAR LIFE tests:

Engine	Transmission	Axle ratio	Adv. HP	0—30 mph	0—60 mph	1/4-mile e.t. @ mph	Est. Net HP
Base 350 engine	A3	3.08:1	300	3.4 sec	8.4 sec	16.1 @ 84	200
Option L46 350	M4	4.11:1	350	2.6	6.4	14.6 @ 97	290
Option L36 427	M4	3.08:1	390	3.4	7.6	15.0 @ 93	280
Option L71 427	M4	4.11:1	435	3.3	7.0	13.9 @ 105	400

The car around which the Carroll Shelby legend was built. Shelby took the ten-year-old A.C. Ace chassis and dropped a high-output Ford V8 into it. The rest is history. This is HPBooks' publisher Bill Fisher's beautiful 289-CID '65. Photos by Tom Monroe.

It's interesting that the snappiest street performance appears to be the L46 hydraulic-cam 350. The 4.11:1 final-drive worked ideally with the close-ratio four-speed and mild cam timing. The L36 427 was handicapped by its 3.08:1 final drive, even with the wide-ratio four-speed. The L71 427 had the same final-drive setup as the L46, but way too much torque to "get out of the hole" with the skinny tires supplied on Corvettes in that day. You had to feather the throttle and slip the clutch like mad to keep from burning up the tires. CAR LIFE testers criticized Corvette engineers for not supplying more rubber as standard equipment on the Tri-Carb 427s. The e.t. potential would certainly have been in the mid-13s with another inch or so of tread width!

But Detroit was still learning.

Cobra interiors were Spartan with few frills. Driving a Cobra was serious business. Photo by Tom Monroe.

THE COBRA CHALLENGE

Perhaps it's not fair to mention the Shelby Cobras in the same breath with production Corvettes. It's like comparing apples and oranges. They were entirely different types of cars, aimed at different markets. Corvettes were somewhat plush, civilized passenger cars that could be used and appreciated by lady secretaries as well as auto enthusiasts.

The Cobra was a car for the purist: a slightly detuned race car. They were barely practical to drive on the street. Cobras had Spartan upholstery, and the skimpy tops would not keep you dry in a rainstorm. Some didn't even have a heater or radio. You had to work at driving a Cobra because of the high-effort controls, such as the race-type quick steering. If a lady secretary wanted one, she'd better be moonlighting as a lady wrestler. It was truly a man's car.

Production volumes of the two cars were dissimilar too. Corvette production in those years was about 20,000 units annually. Shelby made only 1010 cars in five years. Cobras were essentially hand-built custom cars, not *production* cars. At least this was the typical excuse given by the 'Vette owner who had just been blown off by a Cobra.

But the Shelby Cobras were considered production cars

Though early Cobras were fitted with the 260-CID V8, most '63—'65 models had the Shelby-modified *HiPo* 289. Net output was in excess of 200 HP. Photo by Tom Monroe.

Along with the Cunningham, Nash-Healey and Kaiser-Darrin, the Allard J2X represented a post-World War II revival of American sports-car interest. The Allard was available with a Cadillac, Oldsmobile or Chrysler Hemi V8. Photo by Tom Monroe.

Shelby Cobras dominated Corvettes in SCCA B/Production class racing in the early '60s, mostly because of their 800—1000-lb lower weight. Chevy officials argued that the Cobras weren't really production cars.

by race-sanctioning organizations in the 1960s—and that's what counted. Corvettes had to race against them in SCCA A/- and B/Production classes, in FIA International GT classes, and even on the drag strips and at Bonneville.

Besides, it all came back to the same thing I noted earlier: If your 'Vette was blown off by a Cobra at a stoplight, it was little consolation that the Cobra was a limited-production car. Tit for tat.

The tiny, lightweight aluminum-bodied Cobras plagued Corvette racers. The 'Vettes couldn't handle them. Duntov and his crew turned every stone to get the Cobras segregated into limited-production classes, but to no avail. It wasn't until Duntov legalized his L88 and ZL-1 big-block competition options in the late '60s that Corvettes could run on even terms with the 427 Cobras.

But let's look at the Cobra on the street.

First of all, the cars were super light. Carroll Shelby's idea was to drop a muscular American V8 into a small, light British sports car—something on the order of the Hudson-Railton cars of the '30s or the Cadillac-Allards and Chrysler-Allards of the early '50s. Heck, hot-rodders had been dropping small-block Chevy V8s into Austin-Healey 3000s for years.

The A.C. Car Company had this neat little two-seater Ace that used a tubular frame and aluminum body. Even with a huge, long-stroke six-cylinder engine, this A.C. Ace weighed only 2000 lb. It also featured independent suspension, front and rear, on transverse leaf springs. And the car had enough engine setback to get 50/50 weight distribution with a 4- or 5-liter V8. It looked like ideal raw material to Shelby. The A.C. people were glad to sell engine-less cars for him to convert in his California shops.

Many don't know that Shelby first approached Chevrolet to buy small quantities of their small-block V8 to install in his new A.C. Cobra fun car. The GM front office would have none of it—especially as the projected car would be competing with their ultimate image-maker and showroom-traffic builder—the Corvette.

It was then that Shelby turned to Ford—who had just introduced their new thin-wall small-block V8. The T-Bird had long since added two seats and a thousand pounds. Ford had no competing sports models in production. Ford officials apparently saw an easy way to take a poke at the successful Corvette. They were only too glad to supply Shelby with all the engines he wanted.

His first prototype Cobra was completed in October, 1962. It used the 260-CID version of the small-block Ford. The first six cars were used primarily for racing.

Street versions appeared in late 1963 with the 289-CID high-performance engine. These used 3.77:1 final-drive gears and the Warner T-10 four-speed transmission. They weighed only a little more than 2100 lb at the curb. The Goodyear police tires were safe to 130 mph. The initial Cobra engine was completely stock, with exception of the exhaust manifolds. Shelby used a *Tri-Y* system: two cast-iron manifolds on each side connected into a tubular Y that dumped into a head pipe on each side. The undercar exhaust system consisted of straight pipes and simple glass-pack mufflers.

As it turned out, the HP289 engine didn't need much hopping-up to give spectacular performance in the ultra-light Cobra. ROAD & TRACK's test revealed acceleration figures as good as the quickest injected Corvettes:

0—30 mph	2.2 sec
0—60 mph	6.6 sec
1/4-mile e.t.	14.0 sec @ 99 mph
Est. net HP	220

And this was only the beginning. No sooner had the small-block Cobra been established in production and as a consistent race winner in SCCA B/Production and almost every drag-racing and land-speed-record production class it entered, Shelby began another Cobra project. He started

Shelby Cobra 427 was no doubt the best-performing two-seater to come out of the American industry. It was 800-lb lighter than a Corvette, but was a handful to drive. It did move, though! Photo courtesy of Shelby American Club.

Ford 427 engine was a tight fit in the A.C. chassis. Because of this, standard passenger-car exhaust manifolds were used. Most owners removed these in favor of tubular-steel headers. Photo by Steve Christ.

eyeing Ford's 427 NASCAR big-block as a possible way to upgrade performance for all-out International GT road racing.

Actually Ford's 427-wedge engine made some sense for the little Cobra. It was a reasonably compact engine for its displacement, and weighed a little more than 600 lb with an aluminum dual-quad intake manifold. It fit in the Cobra very nicely—though standard FE-engine exhaust manifolds had to be used to clear the frame rails. They were also about 35-lb lighter than the huge cast-iron headers used in 427 passenger cars, but didn't breathe as well.

The big engine also required a redesign of the tubular frame to get additional strength with a minimum of extra weight. And the suspension was completely redesigned with coil springs replacing the transverse leaves all around. The redesign was done by Ford's Kar Kraft personnel, who were flown to the A.C. factory in merry old England.

Amazingly, the redesign added only 400 lb to the little Cobra. Curb weights ran 2500—2550 lb on the 427 street jobs, using the new Ford top-loader four-speed. And 51% of this was still on the rear wheels!

It's difficult to pinpoint reliable acceleration figures for these cars because of the erratic low-gear starts with street tires. It took very delicate throttle feathering and clutch slipping to get just the right amount of wheelspin—and then knowing when to punch the throttle. It took an entirely different technique than driving out of the hole with a 3800-lb passenger car. With a 427 Cobra, timed e.t.s could vary more than 1/2 second from one run to the next.

ROAD & TRACK's figures for a 1965 427 Cobra with 3.54:1 final drive and full street trim seem reasonable:

0—30 mph	2.4 sec
0—60 mph	5.3 sec
1/4-mile e.t.	13.8 sec @ 106 mph
Est. net HP	310

Cobra 427 performance was stellar. How does 0—100—0-mph in 13.8 seconds grab you? Photo by Ron Sessions.

The 310-net-HP rating for the 427 Ford engine may seem low. But you have to remember its restrictive exhaust system. The restriction can't be totally attributed to the standard FE log-type manifolds. Subsequent research has revealed that small-core glasspack mufflers are much more restrictive than many reverse-flow baffle types. The glasspacks make an exciting noise, but they're no good for power. But there wasn't much room for anything else anyway.

Fortunately, a 427 Cobra could whip almost anything on the road even *with* a terrible exhaust system!

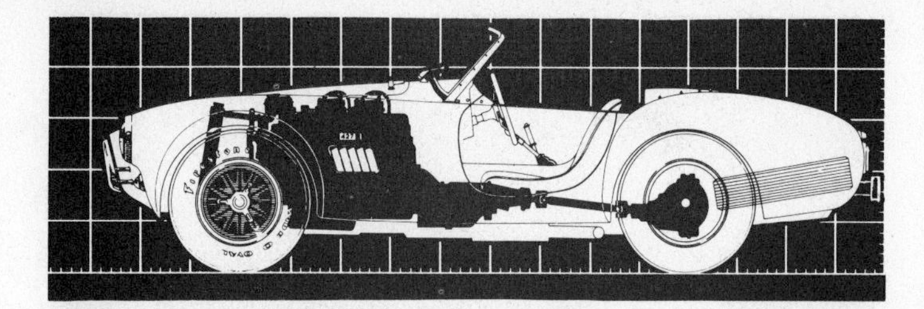

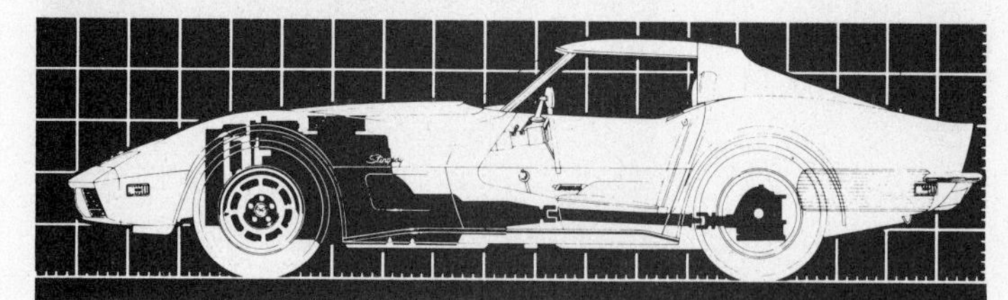

These layout drawings show how the Shelby Cobra (top) and Corvette achieved near 50/50 weight distribution with heavy big-block V8s up front. Engines were set back, with independent suspension in the rear. Drawings courtesy of Road and Track Magazine.

Corvette offered the L88 aluminum heads as optional equipment on 427 high-performance engines in '68—'69. It cost $400 extra to save 70 lb! Photo courtesy of Chevrolet.

DID THEY HANDLE?

So, how did the Cobras compare with Corvettes in general cornering, handling and braking? For sure, the Cobras would usually win in a road race—but a lot of that was superior acceleration and top speed.

Carroll Shelby was the first to admit that the A.C. chassis was out of date when he adapted it for the Cobra. The A.C. was introduced in 1954, and was no world-beater then.

Corvette offered the all-aluminum *ZL-1* 427 engine as a racing option in 1969. It saved 160 lb over the cast-iron 427, and greatly helped Corvette handling by reducing front-end loads.

When Ford engineers helped Shelby adapt the 427 to the Cobra chassis in '65, the front and rear suspensions were totally redesigned and the transverse leaves were replaced with coil-over shocks. The switch helped both handling and ride.

For fast driving on back roads, Cobras were quick, responsive and reasonably predictable in amateur hands. But in all-out road racing against the best GT cars in the world, Shelby himself admitted they were like driving a truck. The Cobra's major virtues were its light weight and the ease with which you could get oversteer by blipping the throttle.

On the other hand, the IRS '63 Corvette was no world-beater in the handling department either. That chassis was designed on a tight budget, and was never anywhere near *state of the art,* even when it was introduced. It had many of the same handling deficiencies the Cobra had—plus it was 800-lb heavier. A major problem with the Corvette IRS was its extreme *bump steer*—the wheels toed in or out with vertical travel. This caused erratic and unpredictable handling. The racer's Band-Aid fix was to limit wheel travel with a very high-rate rear spring. And power-oversteer wasn't as easy to control with the extra weight. To top it off, the higher tire loadings reduced cornering power somewhat.

In braking, I'd have to give the edge to the Cobra, if for no other reason than its lighter weight. Both cars had good four-wheel disc brakes—the 'Vette getting them in '65 with vacuum assist. Given equal brake efficiency, the lighter car will stop quicker. On the road courses, the Cobras could go a little deeper into the corners. Add that to the stronger acceleration out of the corners, and there was no way the 'Vettes could keep up.

But back to the question, did they handle? Let's just say they were both about equally bad!

FOR COMPETITION ONLY

I mentioned earlier that the L88 and ZL-1 Corvettes were competitive with the 427 Cobras in the late '60s. Perhaps I shouldn't be using space on strictly competition

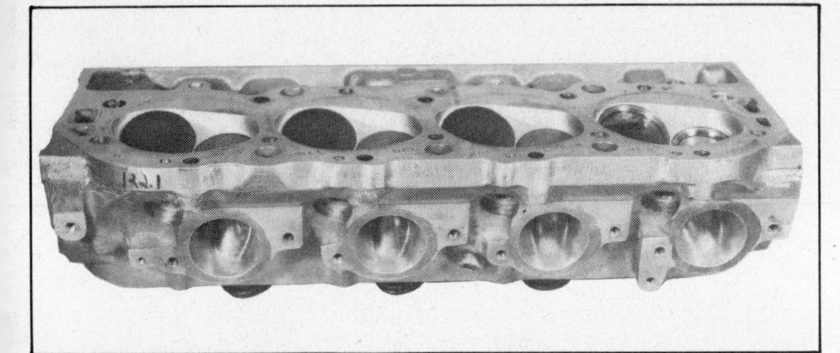

ZL-1 aluminum heads featured recontoured intake ports and *round* exhaust ports—designed to flow best with tubular-steel exhaust headers. Well-tuned ZL-1s developed 585 HP at 6600 rpm in race trim. Photo courtesy of Chevrolet.

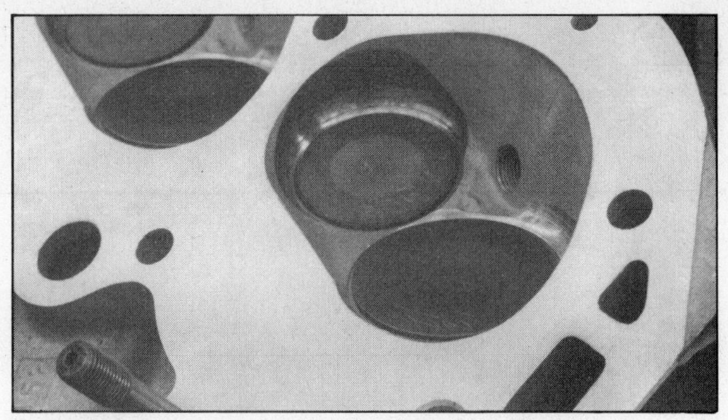

Large, 120-cc open combustion chambers in the ZL-1 heads were later used for all big-block competition heads. They gave good breathing with domed pistons and a 12.5:1 compression ratio. Photo courtesy of Chevrolet.

John Greenwood team out of Detroit was the most-successful "big-time" Corvette racing outfit in the '70s. They built modified GT 'Vettes for racers all over the country. Photo by Lindsay Hoxie.

packages in a book on street machines. But some of this development *did* rub off on the street scene. Of course, no Detroit company ever sold an out-and-out race car that didn't eventually find its way onto the street in some shape or form. Duntov drove a ZL-1 to and from work for years.

Actually, the L88 and ZL-1 options were complete packages available on a base Corvette. These special competition engines came in a stripped car without a radio or heater, with super-duty brakes and suspension, adjustable front/rear brake balance, special Muncie *rock-crusher* four-speed, and other goodies. The cars came through with standard Corvette exhaust systems, which would supposedly be replaced immediately by tubular-steel headers and open pipes. And of course, you would need special tires to do any serious racing. But other than these two areas, it was expected that an L88 or ZL-1 Corvette could be raced right off the assembly line!

The engines were especially interesting. The '67—'68 L88 used the L71 four-bolt-main block with aluminum heads (similar to the cast-iron heads), 12.5:1 pistons, a big 830-cfm Holley double-pumper carb on the high-riser manifold, and a special solid-lifter cam with a beefed-up valve train. Also, the *Magnafluxed* rods used floating pins and larger, 7/16-in. bolts. A high-volume oil pump was used. The bottom end was built to allow sustained 6800 rpm. High compression required the use of super-premium fuel with a minimum Research-octane rating of 103—still widely available in the late '60s.

The ZL-1 engine of 1969 was even more sophisticated. The aluminum heads were redesigned with round exhaust ports. Large, open combustion chambers improved combustion and breathing with reduced shrouding around the valves. Weight was dramatically chopped with a new aluminum cylinder block with iron liners. Altogether, the ZL-1 weighed 160-lb less than a standard L36.

The ZL-1 cam and valve train were further upgraded to give more top end and cure the common big-block problem of valve-spring breakage. The bottom end was firmed out with stronger rod bolts and a still-better lubrication system. Duntov felt confident in raising the redline on this monster to an incredible 7600 rpm.

According to Chevy engineers, a well-tuned ZL-1 with tubular-steel exhaust headers and open pipes—in race trim—should deliver *585* big ones at 6600 rpm, with a whopping 520 ft-lb of torque at 5000 rpm. That's well over one horsepower for every pound of engine weight—and 1.38 horsepower for every cubic inch!

The actual acceleration of an assembly-line ZL-1 Corvette was probably as good as a competition 427 Cobra—for the simple reason that a ZL-1 engine put out about 100 more horsepower than a competition-tuned 427 Ford. When you combine this with the huge weight saving with the aluminum block and heads, the power-to-weight ratios are similar—the Corvette might even be *better.* I know an assembly-line ZL-1 Corvette weighed about 2950 lb with five gallons of gas. Use that weight with the claimed 585 HP and you come up with some pretty impressive acceleration figures.

In fact, I have a standing-start quarter-mile time from Chevrolet Engineering of 11.0-second e.t. at 127 mph. That sounds reasonable, with the weight and horsepower involved. The best I ever heard for a competition 427 Cobra was 11.4-second e.t. at 123 mph. Arguments form on the right!

Wild graphics on the Pontiac Trans-Ams of the late-'70s helped car buyers forget declining performance under the hood. Photo by Tom Monroe.

When GM lifted its ban on over-400-CID engines for 1970, Olds dropped the 455 V8 into the 4-4-2. Olds didn't need Hurst's help in '71—'72. Photo courtesy of Oldsmobile.

10

Supercars in the '70s

Choking the golden goose

1970: THE PEAK OF SUPERCAR DEVELOPMENT

Theoretically, the 1970 factory supercars should have been the hottest and most-exciting examples of the breed. Federal exhaust-emission regulations had not yet taken a big bite out of performance. You could get 100+ octane gas at every corner filling station for less than 40¢ a gallon. The GM front office had lifted their ban on over-400-CID engines in intermediate-size cars. Compression ratios had not yet started to come down for unleaded gas. Low-profile belted tires had given a big improvement in traction for everyday street cars. Many companies were doing extensive cylinder-head development on airflow test benches. Performance engineering had evolved into a high science. The stage was set for the wildest crop of supercars ever.

But when you study the actual acceleration test figures on these 1970 cars, the picture gets very spotty. Some of the cars peaked out. Some improved a little from late-'60s performance. And a few actually started to decline in performance, even though superficial specs on cubic inches and weight suggested something better.

It's almost as if the seeds of destruction of the Detroit supercar had already been planted in 1970. The design engineers didn't seem to have the old punch. They seemed to be reading the handwriting on the wall more than the blueprints.

One of the first bad signs was escalating insurance rates on high-horsepower cars. This effect took an immediate bite out of the market, especially with high-risk young drivers who comprised a major percentage of it. The Muskie Clean Air Act was enacted that year, promising ever-tightening federal exhaust-emission standards for the next decade. More federal safety standards meant additional weight and cost. And most important, traffic-safety officials were openly criticizing these high-performance cars with 140-mph top-speed potential.

Detroit manufacturers were not exactly *gung-ho* when they introduced their 1970 supercars. Or maybe I should say the designers took off their boxing gloves and put on their white gloves. There were a number of examples where the major development for 1970 seemed to be aimed more at *civilizing* the cars than getting maximum performance and excitement. Performance was not *downgraded,* but the new aim seemed to be to deliver the performance with less fuss and commotion. If these compromises soaked up a little acceleration, that's the way it was.

Examples:

Chevrolet eliminated the hood bubble on the Corvette. This required a new low-riser intake manifold for the big-block high-performance engine. The low-riser manifold cost 10—15 HP compared with the earlier high-riser manifold. Also, to pass corporate noise standards, a snorkel-type closed air cleaner was adopted on the Chevelle. It was so restrictive that it cost another 15—20 HP compared with the big 14-in. open-element filter. Result: The '70 454-CID LS-6 engine produced less horsepower than the smaller, '69 427-CID L72—even though both used the same heads, cam, carburetor, exhaust system, and so on.

Above: Chevy used the 454-CID big-block in SS Chevelles in the early '70s to offset gradual weight increases. But performance was not much better than the 396s of the late '60s. Note *Cowl Induction* with the vacuum-operated air-intake flap. Photo courtesy of Chevrolet.

Due to hood-clearance problems on '70-model bodies, Chevrolet had to use a new *low-riser* intake manifold on 454-CID high-performance engines. Its sharper corners cut 10—15 HP at the top end.

1971 Olds 4-4-2 convertible had the W30 option, providing dual hood scoops and long-duration cam. By this time, the W31 had 455 cubic inches to help move the weight. Photo courtesy of Oldsmobile.

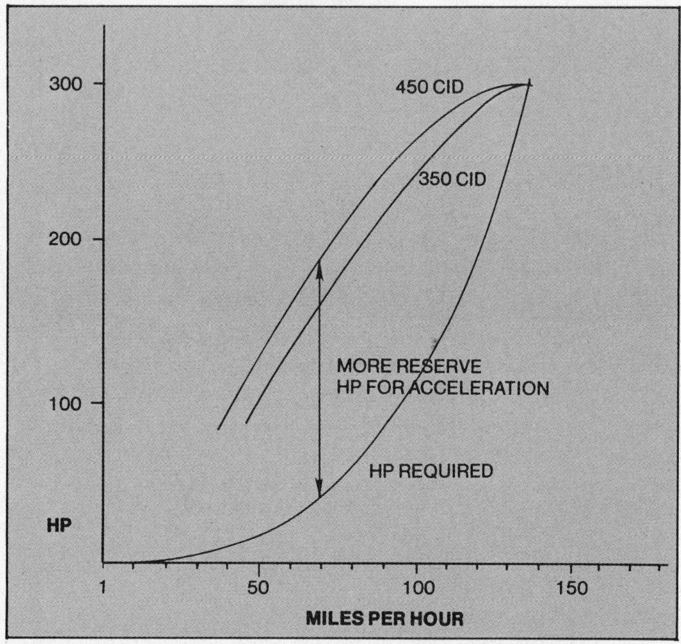

Chart shows why the companies switched to bigger and bigger engines in the '70s. Emission controls were beginning to reduce specific output, and weight was creeping up. Simply put, bigger displacement meant more torque at lower engine speeds.

Oldsmobile eliminated their wild under-bumper air scoops for the W30 and W31 engines in favor of more-conventional scoops in a special fiberglass hood. Putting the cold-air system in the hood was less expensive and easier to install on the assembly line. It also allowed the use of a vacuum-operated trapdoor to feed warm air to the carburetor during warmup and at part-throttle cruise. That, in turn, permitted leaner primary carburetion and better gas mileage. But the price was less ram pressure, and maybe a loss of 20 HP above 70 or 80 mph. The increase in displacement from 400 to 455 cubic inches for the 4-4-2 barely made up the difference.

Chrysler switched to a hydraulic camshaft for the wild Street Hemi engine. It had the same timing and lift as the previous solid-lifter cam, so performance was theoretically the same. But you lost a few-hundred rpm off the top of the rev range. The big advantage of the hydraulic cam was that valve clearance didn't change—and valve timing and lift didn't gradually deteriorate as the miles built up.

In these examples, performance wasn't particularly helped by changes aimed at civilizing the wilder supercar combinations of the late '60s. But there were some cases were both performance *and driveability* were helped by developments that would not normally be considered hop-up engineering.

Probably the best example was the Z/28 Camaro. SCCA changed their Trans-Am rules to allow standard blocks in homologated cars for the 1970 season. Chevrolet took advantage of this in the Z/28 by replacing the 302 with the 350-CID small-block. But instead of trying to turn all those extra cubes into wild horses, they went to a *milder* camshaft at the same time. The result was a good increase in power at the top end—and a big improvement in low-speed throttle response and mid-range torque. Acceleration improved in every speed range: The car was much-more pleasant and responsive to drive on the street. This was a perfect example of the white-glove type of engineering that was beginning to show up around 1970.

This was further illustrated in the next couple of years by *the big-inch syndrome.*

THE BIG-INCH SYNDROME

Pontiac's orginal concept of the *civilized* supercar was a big-block V8 of 400 CID or less in an intermediate-size coupe. That's right, the GTO. It's history that Pontiac's concept was embraced by other GM divisions and other Detroit manufacturers. The GM front office reinforced this concept by clamping an official lid of 400 CID on *any* intermediate model in the late '60s. It seemed like good public relations aimed at the safety critics, if nothing more.

Chrysler was the first to rock the boat when they adopted their 440-CID high-deck B-block for new Dodge and Plymouth supercars in 1967. Those extra cubic inches definitely gave an edge in overall street flexibility. Ford followed in '68 and '69 with the 428 Cobra Jet engine for Mustangs, Cougars, Torinos and Cyclones. That put the handwriting on the wall for GM. With their intermediate

Dodge made the Super Bee performance package available on the restyled the Charger in 1971. Previous 383 B-block was bumped to 400 cubes and more carburetion was used. Photo by David Gooley.

1970 Buick GS 455 Stage-1 was perhaps the most-potent Buick ever produced.

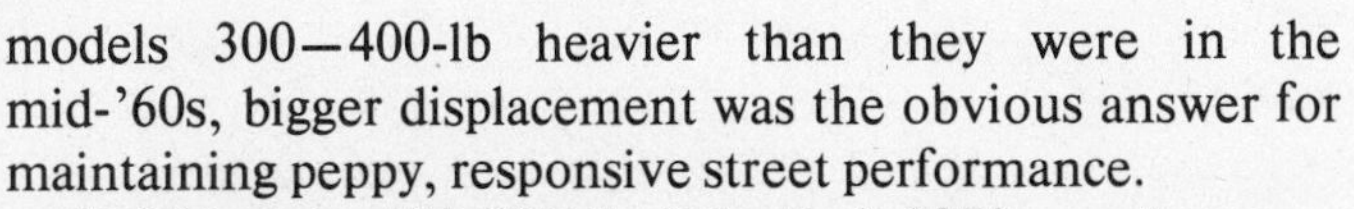
models 300—400-lb heavier than they were in the mid-'60s, bigger displacement was the obvious answer for maintaining peppy, responsive street performance.

GM lifted the 400-CID limit for their 1970 muscle cars.

In previous paragraphs, I mentioned how some of these later big-displacement cars sacrificed a little performance for practical considerations like low-speed response, reduced noise, easy maintenance, manufacturing convenience—and so on. The die-hard performance addicts weren't calling all the shots on supercar design any more. Factory engineers seemed determined to use bigger displacements to make these high-performance cars smoother, sweeter and more driveable than previous performance cars.

Nevertheless, there *were* some great engines that came out of this period in the early '70s. To mention only a few:

Buick 455 Stage-I—When Buick upped the displacement of their Stage-I engine from 400- to 455-cubic inches, they also increased valve sizes and used a cam with more duration and lift. The 455 Stage-I combination was definitely better than the 400-CID version—not so much in brute horsepower, but in everyday driveability. The 455 Stage-I was Buick's crowning glory in street engines.

Chevrolet 454 LS-6—Earlier, I criticized the top-end performance of the '70 LS-6 because of a switch in intake manifolds and air cleaners. The full potential of the engine was never realized. But don't get the idea that it wasn't a gutsy engine to drive on the street. The low-end torque was something else. The inferior manifold and air cleaner weren't felt below 4000 rpm. In everyday driving, this was one of Chevrolet's sweetest.

Ford 429 Cobra Jet—Keep in mind that this was the new *385-series* luxury engine introduced in '68. It was an entirely different design than the 427 and 428 *FE* engines of the '60s. Canted valves, huge ports and open combustion chambers gave outstanding power and torque throughout the speed range, even in standard passenger-car form.

Ford engineers went a step further for the '70—'71 Cobra Jet version by using special big-port heads, high-riser intake manifold, Rochester Quadrajet carb, high-output hydraulic-lifter cam and better-breathing exhaust manifolds. These helped the top-end performance, but not overall driveability.

Though brute power was down a bit from previous big-blocks, the '72 'Vette 454-CID LS-6 was one of the sweetest-running street engines ever. Photo courtesy of Chevrolet.

The 370-HP 429 CJs certainly had more potential than the earlier 428 CJs. But only a handful were ever built, and Ford did no further development work after the initial design. Parts were always difficult to obtain, and some Ford dealers worried about future service problems. Color the 429 CJ exciting—but questionable!

Ford 429 Super Cobra Jet—Another performance version of the 385-series engine was the 429 SCJ. Also introduced in '70, the SCJ used the Cobra Jet cast-iron heads, but with a higher-output cam and mechanical lifters. A similar high-riser intake manifold was installed, but with a different bolt flange to mount the Holley 4160 carb. For high-rpm durability, the block featured four-bolt mains. The pistons were forged. Compression was a hefty 11.3:1. There was also a factory oil cooler. The engine was conservatively rated at 375 HP to make it competitive in certain drag-racing classes.

The 429 SCJ was dropped along with the 429 CJ after the

Ford's *385-series* big-displacement V8, introduced in '68, had staggered valves and a good-breathing port layout. It produced lots of horsepower per cubic inch. Performance models used a 429-CID version. Photo courtesy of Ford Motor Co.

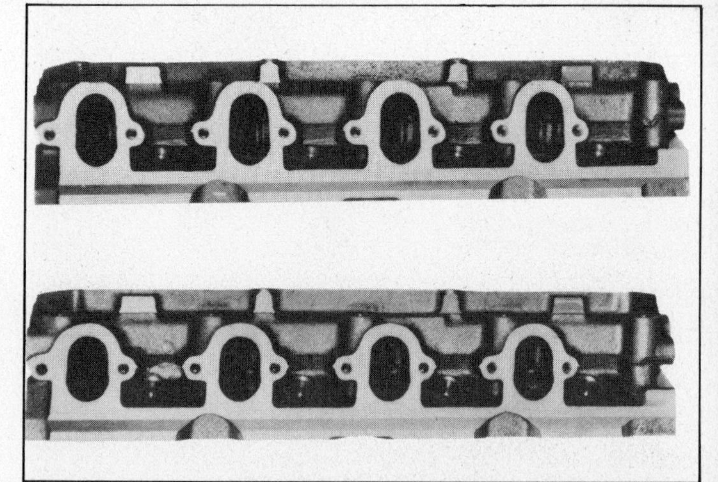

Special big-port heads on 429 Cobra Jet engines, at top, produced an extra 20—30 HP over standard heads. Very few of these engines were sold to the public, and little development was done.

'71-model run: victims of poor sales, high insurance rates and a move towards low-lead, low-octane fuel.

Oldsmobile 455 W30—Earlier, I criticized the '69 400-CID W30 for its driveability problems. In my opinion, the factory camshaft was perhaps *too radical* for the long stroke of the '69 400-CID Olds. The cam had 328° duration and 0.474-in. lift.

But those extra 55 cubes in 1970 made an entirely new car out of the W30. With the identical camshaft, the 455 had a better low end, better mid-range, smoother idle, better fuel economy. It also had a few more horsepower at the top end. It was a beautiful combination.

Interestingly enough, the wild 328° cam was only used with four-speed manual transmissions. Olds specified a milder 286° or 294° cam with W30 automatics. This was a wise move for two reasons: First, the automatic-transmission modulator needed a strong intake-manifold-vacuum signal to time shifts and regulate hydraulic pressure to the clutch packs. And second, Olds figured that the die-hard performance fans would buy the four-speeds, and those opting for the automatic transmission might appreciate a more-flexible low end. The sacrifice of 10—15 HP at the top end paid off handsomely in everyday driveability.

Pontiac 455 H.O.—This one didn't come along until 1971, so it was saddled with a low compression ratio of 8.4:1. The 455 H.O. used a fairly mild camshaft with 288° duration and 0.420-in. lift. It was a grind that had been used on 421 and 428 H.O. engines back in the '60s. But all this was balanced by adapting the excellent big-port heads, high-riser intake manifold and split-flow exhaust headers from the wild Ram Air IV engine of the '69—'70 period.

The combination of mild camming and compression with larger porting and manifolding gave an engine with a smooth low end and broad torque curve between 2000 and 4000 rpm. Pontiac clamped a tight redline of 5200 rpm on the 455 H.O. because of its long stroke, large-diameter main bearings and cast-iron connecting rods. It wasn't a

Oldsmobile replaced their famous under-bumper air scoops with more-conventional twin hood scoops in 1970. The hood scoops looked more impressive, but were less efficient in ramming cold air to the carburetor. Photo courtesy of Oldsmobile.

GTO 455 H.O. of '71 was a real stormer. Photo by Ron Sessions.

In the Firebird Trans-Am, the 455 H.O. sported a *shaker* hood scoop. Photo by Ron Sessions.

Pontiac upgraded the big 455-CID engine for high-performance street use in 1971. This 455 H.O. of the '71—'72 period used big-port heads, and intake and exhaust manifolds from the Ram Air IV 400.

NASCAR speedway racing boomed in the '70s. Even though Detroit was backing off on performance, they still offered over-the-counter goodies. This is Richard Petty.

rev-happy engine like some. But it put 300 net HP and 455 ft-lb of torque into a Turbo Hydra-matic 400 transmission. It was one of Pontiac's nicest street engines ever.

DID THEY RUN?

As pointed out earlier, I wouldn't say that these late big-displacement supercars showed any consistent performance advantage over their 400-CID counterparts of the late '60s. Standing-start quarter-mile elapsed times remained in the mid-to-high 14s at 95—98-mph trap speed, with 0—60-mph times in the 6-second bracket.

There were various reasons: Where power went up with the extra cubes—like the Stage-I Buick and W30 Olds—higher curb weights offset the gain. In some cases, like the LS-6 Chevy and 455 H.O. Pontiac, there was no increase in top-end power over smaller-displacement, more-highly tuned engines of the late '60s. The 455 H.O. Pontiac, for instance, developed 20—30-fewer net horsepower than the 400-CID Ram Air IV engine.

And then there's the Ford 429 Cobra Jet. Though it had lots of potential, the engine was never developed to deliver consistently strong street performance *as it came off the assembly line.* You could tune it to run strong. But few owners—and fewer Ford mechanics—had the know-how. It was a giant that never got a chance to show its muscle.

Where these later big-displacement performance engines really excelled was in peppy, responsive street driving—especially with automatic transmissons. The extra displacement and generally mild camming combined to give these engines a real wallop at low and medium speeds. And the move to lower compression ratios in '71—'72 didn't seem to hurt them much. These were just really nice street engines—for an enthusiast or Aunt Jane.

But in all-out acceleration, they were no quicker than smaller-displacement supercars of the late '60s.

EMISSION RESTRICTIONS CLOSE IN

Historians will undoubtedly argue the reasons behind the public panic on air pollution in the '60s. Certainly, there was a serious smog problem in the Los Angeles basin and a few other areas of the U.S. And a rising environmental movement was demanding government action. It was probably inevitable that the giant auto industry would be hit with restrictions on exhaust emissions.

But no one was quite prepared for the *Muskie Clean Air Act* of 1970. That legislation was rushed through with practically no hard facts on the relationship between auto emissions and air quality. Admittedly, the legislators guessed. So, many figures were guesstimated in smoke-filled committee rooms. The auto industry was suddenly faced with the need to cut carbon-monoxide, hydrocarbon and oxides-of-nitrogen emissions by more than 90% in a 10-year period. It was law.

Reductions in allowable emissions were to come in gradual steps every two or three years. The first standards applied only to carbon monoxide (CO) and hydrocarbons (HC). Limits on oxides of nitrogen (NOx) were to come in 1973, and a big cut in all three pollutants was set for 1975.

Industry engineers were in an overnight panic. Early CO and HC standards were no big problem, and could be met with minor changes in carburetion and spark timing—sometimes with an air-injection pump. NOx standards set for '73 could probably be met by recirculating exhaust gas into the intake manifold to keep down combustion temperatures. Of course, this would take a toll in power and fuel economy. There's no way you can maintain performance efficiency by diluting the incoming air/fuel mixture with exhaust gas. But at least the engineers felt the NOx standards could be met with simple changes in intake-manifold castings and a vacuum-operated exhaust-gas-recirculation (EGR) valve.

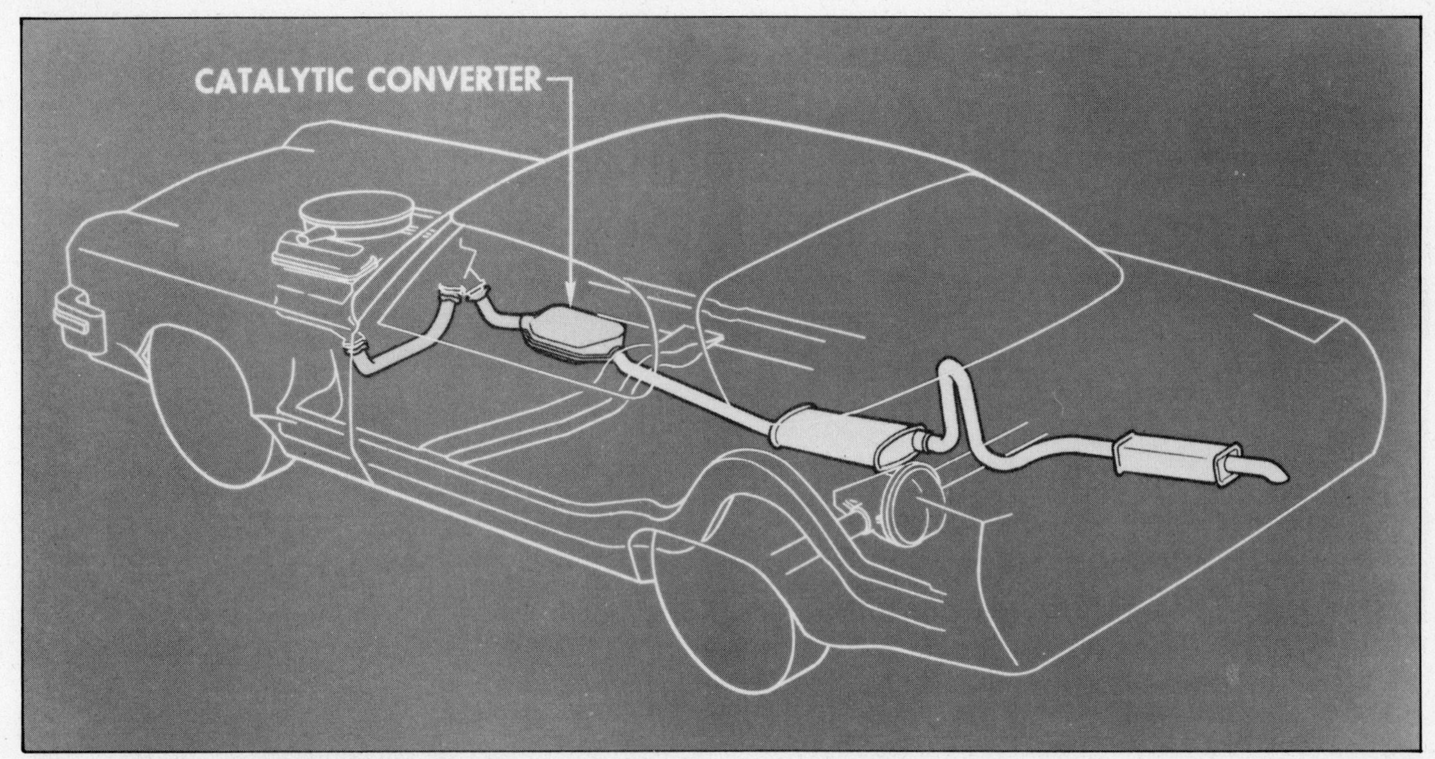

Tightening federal exhaust-emission restrictions forced the industry to use catalytic converters in 1975. Converters caused a big restriction in exhaust flow, choking off horsepower. Drawing courtesy of General Motors.

LT-1 Corvette engine of 1970 was the peak of small-block V8 factory-performance development. It gave a net output of 320 HP from 350 cubic inches; turned up to 6500 rpm with no fuss. Photo courtesy of Chevrolet.

The big reductions in allowable pollutants scheduled for 1975 were another story. All U.S. automakers agreed that meeting these standards would require something exotic like a *catalytic converter.*

In case you didn't already know, a catalytic converter is a muffler-shaped item in the exhaust system. It contains a precious-metal *catalyst* agent—usually platinum or palladium—in a honeycomb substrate or pellet bed. When exhaust gas passes through, a chemical reaction converts HC and CO into carbon dioxide and steam.

Such a device adds weight and cost to the vehicle—and generates a lot of exhaust back pressure that hurts performance. Also, all known catalyst materials would be contaminated by leaded gasoline, neutralizing the affect of the converter. This meant the oil industry would have to offer lead-free gasoline by 1975.

It was at this time that GM, and eventually the rest of the industry, decided to switch to low compression ratios on '71 and '72 models in preparation for the coming unleaded gas. Without tetraethyl lead to slow the combustion process, high compression would cause damaging detonation in these supercar engines.

All around, it was a sad time for performance enthusiasts. The future suddenly looked pretty bleak. And remember, this was before any hint of the Arab-oil embargo of 1973 that sent gas prices skyrocketing. Everything seemed to work together after 1970 to kill the supercar deader than a mackerel. It all happened within the space of three or four years.

Perhaps the most-effective commentary I can make on the whole horrible process is to trace the *de-evolution* of one popular high-performance engine during this period. For instance, look at the top 350-CID small-block option for the Corvette:

1970: This marked the introduction of the fabled *LT-1* combination. It was the peak of factory performance development of the small-block V8 engine. Practically every off-the-shelf goody was used. Cost and complexity were considered secondary to performance. The 11:1 compression

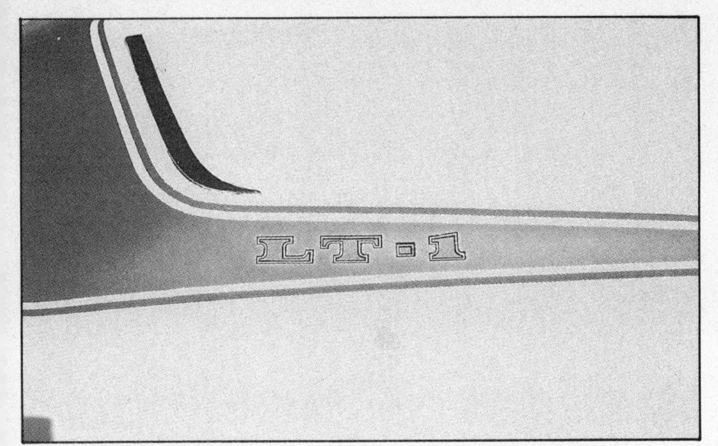

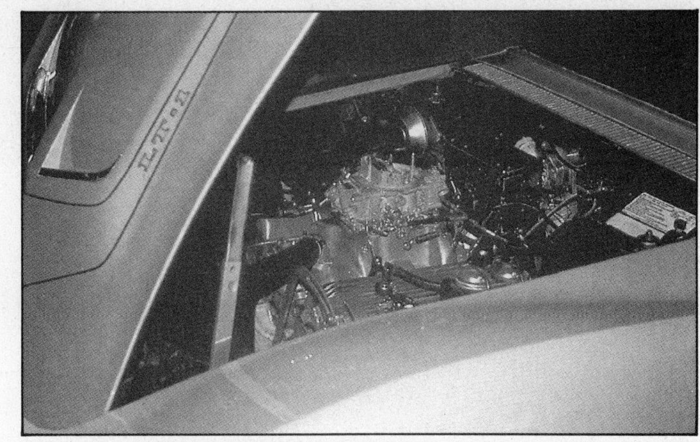

LT-1 Corvettes of the early '70s offered a good blend of performance and balanced handling. Photos by Tom Monroe and Ron Sessions.

ratio required 100+ octane gas, without apology. Net power as installed in the car: 320 HP at 5600 rpm.

1971: Compression ratio was reduced to 9.0:1 for unleaded 91-octane gas. Factory net power: 280 HP at 5600 rpm. Total spark advance was set way back.

1973: The first NOx regulations required extensive reworking of the intake manifold to allow exhaust gas to be introduced and distributed into the air/fuel mixture. Chevrolet was busy meeting government regulations. Consequently, they couldn't afford the money and manpower to re-engineer the excellent LT-1 high-riser aluminum manifold for EGR. After all, the LT-1 represented a small percentage of total sales. This meant going back to a standard low-riser cast-iron manifold. Using the low-riser meant reverting to the bread-and-butter Quadrajet carburetor, rather than the 780-cfm Holley 4-barrel—which didn't fit the standard manifold. So the '73 engine was not only hit by the smothering effects of EGR—it got inferior carburetion and manifolding in the bargain.

And there's more. Tighter HC and CO standards in '73 couldn't be met with the LT-1 solid-lifter camshaft. This meant using the hydraulic-lifter street cam of the late '60s, resulting in another power cut above 4000 rpm. The '73-engine changes were so extensive that a new *L82* option code was adopted. The honored LT-1 specificiation was quietly buried. Factory net rating for the new L82: 250 HP at 5200 rpm—70-HP less than the 1970 models.

1975: This was the year of the extreme cuts in allowable exhaust emissions. Catalytic converters were almost a necessity on all cars, big and small. GM chose a pellet-bed converter, in contrast to the *honeycomb* type used on most Ford and Chrysler models. The pellet type was considerably cheaper, but also more restrictive.

Corvette engineers wanted to run two converters with the L82 engine. Two would theoretically cut exhaust back pressure by 75%. But the catalyst pellets didn't get hot enough to *light off* the exhaust gasses with two converters. So Chevrolet ended up bottlenecking the exhaust through one converter—and trying to gain a little by using two pipes and small mufflers at the back. Net power: 205 HP at 4800 rpm.

It doesn't take much to figure what this 36% reduction in net power did to the Corvette's acceleration in the early '70s. When you add the increased weight of 5-mph bumpers and necessary reinforcements needed to meet new crash standards, performance really took it on the chin. How about 2-seconds more e.t. and nearly 15-mph slower trap speed in the quarter-mile? Like I said, it was a sad time for die-hard performance fans.

Pontiac made the last serious stab at street performance in the '73—'74 period with the fabled 455 Super Duty engine. It netted 310 HP with all emission equipment, and had a beefed-up bottom end to turn 6300 rpm. Photo courtesy of Pontiac.

PONTIAC THROWS THE LAST PUNCH

Auto historians are still trying to figure out how the Pontiac *Super Duty 455* engine of the '73—'74 period ever happened. Certainly all odds were against it. Allowable limits for HC and CO emissions had just been cut, and the first NOx standards were imposed on '73 models. Gasoline octane was beginning its long decline from the lush days of the late '60s. Insurance rates on high-performance cars were going up and up. All the other Detroit carmakers were backing off radically on performance development. With all the hassle on emissions, there wasn't even enough manpower to do bread-and-butter engineering.

Then all of a sudden appears this exotic performance

Supercar swan song. Only about 1200 or so of these beautiful Super Duty 455 Trans-Ams were built in '73—'74. Photo courtesy of General Motors.

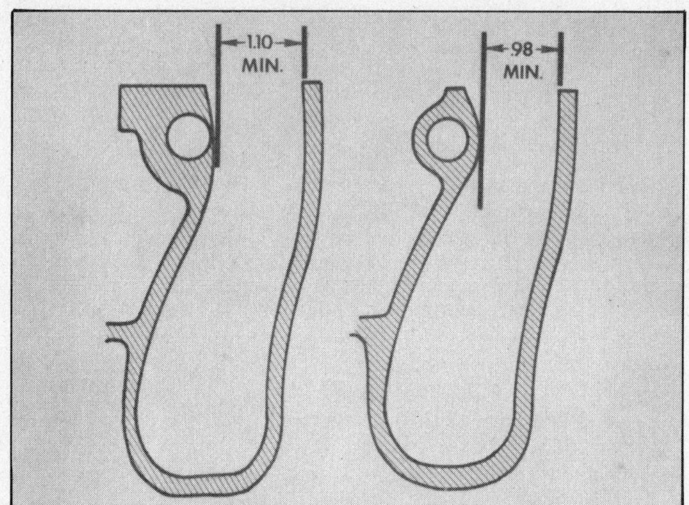

Intake ports were widened on the Pontiac SD 455 by inserting steel tubes in the pushrod holes, and opening the casting wall out to the edge of the tube. Port width was increased 1/8 in., left. Drawing courtesy of Pontiac.

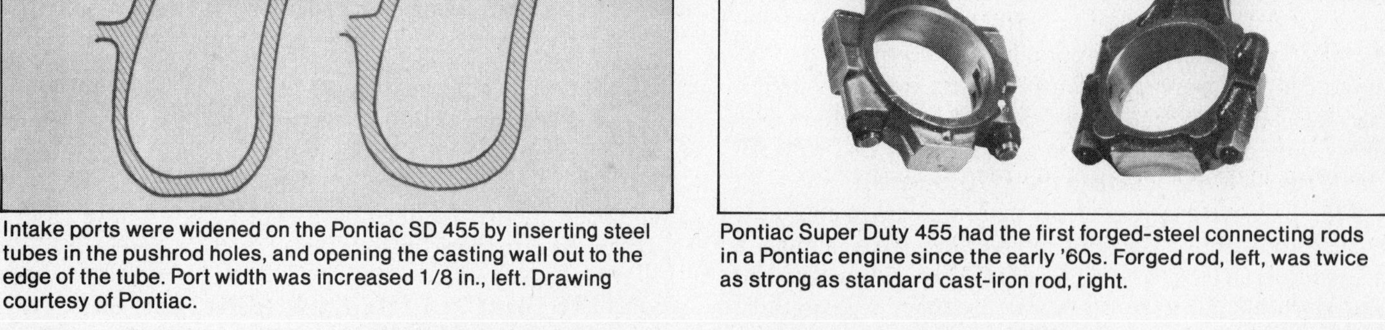

Pontiac Super Duty 455 had the first forged-steel connecting rods in a Pontiac engine since the early '60s. Forged rod, left, was twice as strong as standard cast-iron rod, right.

engine out of Pontiac Engineering—practically an all-new design. It had dozens of new special-design parts, the result of hundreds of hours on the dyno, with specs that read like a page out of the '60s. Enthusiasts were dumbfounded. Why? How? We were tickled to see another thrust at real performance. But how could such an extensive project ever get by the GM front office?

It's generally agreed that two prestigious Pontiac engineers—Herb Adams and Tom Nell—pushed the thing to completion. A sympathetic general manager apparently looked the other way. The boys were even able to certify the engine to '73 emission specs. They did it by using a Quadrajet carburetor, EGR, and a Ram Air IV camshaft in conjunction with standard rocker arms, which cut valve lift to 0.470 in. It was touch and go on the emissions, but the engine finally passed.

Actually, the net output in standard form wasn't all that impressive. An honest 310 HP at 4400 rpm was all it could manage with its low compression, EGR and moderate valve lift. The restrictive crossflow muffler in the Firebird chassis didn't help either.

But the guy who wanted to make a few minor mods on

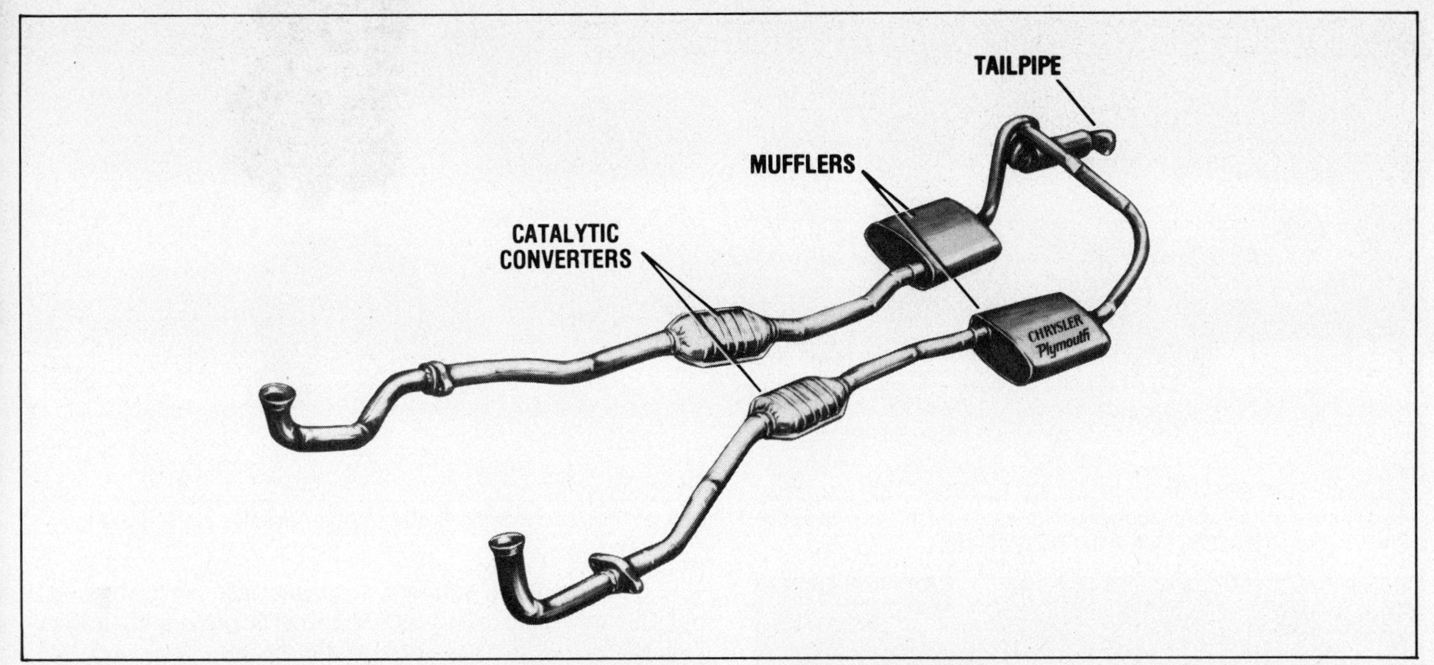

Chrysler tried twin catalytic converters on some performance models in the late '70s. They added 10—15 HP, but were expensive, and emissions were high on cold starts. Drawing courtesy of Chrysler Corp.

his SD 455 had a tremendous base to start from. The cylinder heads alone far surpassed anything Pontiac had ever done. SD 455 intake ports were made 1/8-in. wider by sleeving the pushrod holes with steel tubes and moving the port wall right out to the edge of the tube. Intake-port height was increased 1/8 in. by recoring the port interior. Exhaust ports were also recountoured into more of a venturi shape. Flow was said to improve 20% on both intake and exhaust.

Pontiac engineers worked on the bottom end, too. The rev limit went up from the stock 5200—5500 rpm to at least 6200. Remember that earlier 455 H.O. Pontiacs were notorious *lazy revvers.* The cylinder-block casting was extensively reinforced around the bulkheads, with four-bolt bearing caps on *all five* mains. Pontiac didn't OK a forged-steel crankshaft—but they did tool up their first forged-steel connecting rods in a decade. They even made the rods from high-quality chrome-nickel-alloy steel. In conjunction with hard main and rod bearings and a high-volume oil pump, the engineers figured that the bottom end would be safe to deliver up to 600 HP at 6000 rpm.

It can be surmised from the 310-HP net output that acceleration figures with a 3700-lb Firebird were not earth-shaking. Maybe one could expect a 1/4-mile e.t. in the high 14s at 95 mph. The beauty of the Super Duty 455 engine was what you could do with it. It was a solid base to add more compression, carburetion, camming and a set of tubular-steel headers. Several owners drag raced the combination with some success in NHRA Super-Stock classes.

Unfortunately, Pontiac never made enough of them to gain much of a reputation. Sources vary on the exact number, but apparently only 1238 cars went out with the SD 455 engine in the 15 or so months it was available. I believe they were all Firebirds.

CARRYING ON

After catalytic converters descended on performance fans in 1975, there were several notable cases where company engineers refused to give up on high performance in the face of tightening emission standards. Some did really outstanding performance work on production models.

For instance, Chrysler was able to hold out for two years against converters on their 360-CID 4-barrel *police* engine by using extremely lean primary carburetion with a belt-driven air-injection pump. They got 220 net HP at 4400 rpm with a neat dual exhaust system—and met all emission standards. Then, when tighter 1978 standards definitely forced the converter issue, they successfully used *two* converters, and got 195 HP at 4000 rpm. This dual-converter engine wasn't continued very long because of the cost of the two units. But I commend Chrysler engineers for making the idea work.

Corvette engineers also did some clever hop-up work on the L82 engine in the late '70s. Remember, its net output dropped to 205 HP when the converter was first used in '75. This output was raised to 220 HP at 5200 rpm in the next four years, largely through detail refinements in the exhaust and intake systems. Remember, this engine was still wheezing through a single pellet-type converter. At the same time, the weight of the Corvette was dropped 200—300 lb by judicious use of aluminum, fiberglass and thinner-gage steel. So by 1980, acceleration of the L82 version was virtually equal to the first 1973 model, which had 250 net horsepower. Considering the constraints Chevy had to work under, it was performance engineering at its very best.

Pontiac didn't give up on Trans-Am performance either. After phasing-out their 455-CID long-stroke engine in

Head Pontiac stylist, John Schinella, did wonderful things with spoilers, flares and decals—to keep Firebird Trans-Am sales at the top of the industry though the '70s. This is the 1976 model.

Pontiac's John Schinella came up with exciting body graphics for the Firebird Trans-Am. Here is a '77 T/A 6.6 hood. Photo by Ron Sessions.

1976, they worked diligently with the 400-CID version. They upped net output from 200 to 220 HP with a high-output cam and a new dual exhaust system behind a catalytic converter similar to Corvette's. Incidentally, this same system was also adopted on the Z/28 Camaro. Then, when the 400 engine was phased out after the '78-model run, Pontiac engineers got busy and turbocharged their 301-CID short-stroke V8—and still got 200 HP!

These horsepower figures sound pretty tame after the excesses of the '60s. But let me assure you it took tremendous engineering to get them—and still meet federal emission and fuel-economy standards of the late '70s. Performance fans rewarded the efforts by buying and street-racing the late models with as much gusto as ever.

A SUBSTITUTE FOR PERFORMANCE?

With emission and fuel-economy regulations closing in on high-performance engines in the '70s, Detroit sales people had little choice but to look for other ways to attract a still-boiling youth market.

One fad that was quite successful in the mid-'70s was the use of *tack-on body trim.* Sure, there were many cases of wild body graphics, spoilers, scoops and so on in the '60s, but the new approach was somewhat more sophisticated. In addition to all that other stuff, the new trim included fender flares and air dams. It reflected new emphasis on handling, braking and slick aerodynamics—emulating road racing more than drag racing.

The apparent father, and certainly the most-skillful practitioner of the new art, was John Schinella—who was appointed director of the Firebird styling section in 1973. He was the main thrust behind the wild fender flares, spoilers, scoops and "screaming chicken" hood decals that graced the Trans-Am models of the late '70s. Schinella was an absolute master in jazzing up the styling and image of a standard body with a few pieces of plastic that could be quickly attached on the production line. This tack-on trim was certainly a major factor in the sales success of the Firebird Trans-Am. It sold in excess of 90,000 units a year in the late '70s. I touched on this development in an earlier chapter, but I can't overemphasize its importance in the evolution of the youth/performance car.

John Schinella's flamboyant styling themes made street fans forget the 14-second e.t.s and 400-HP engines of the '60s. They bought his 200-HP, 16-second Trans-Am with as much enthusiasm as they ever showed for the GTO. And it's likely that the Pontiac division made a better profit margin on the Trans-Ams, because of the emphasis on expensive options such as T-Roofs, tilt-wheels and multiplex stereo radios. And there was certainly less warranty trouble with broken engines, transmissions and rear axles! Decals and fender flares have no moving parts.

Needless to say, the Trans-Am concept of tack-on body trim attracted a lot of copiers in the mid- and late '70s. But no manufacturer approached Schinella's deft touch. Chevrolet tried it with the Z/28 Camaro, but couldn't approach mid-'70s Trans-Am sales figures. In fact, Chevy dropped the Z/28 for two years after catalytic converters came in, mistakenly calculating that performance was the only major factor in selling this type of car. Pontiac kept plugging away with the Trans-Am. By the time Chevy brought back the Z/28 in 1977, they had lost another big chunk of the market.

Chevy stylists picked up on the Trans-Am flamboyance with the Z/28 Camaro in the '70s. This is the 1981 Z/28. The Z/28 didn't catch up to Trans-Am sales until the early '80s. Photo courtesy of Chevrolet.

Ford and Chrysler tried to capture the Trans-Am mystique with more emphasis on stripes, decals, spoilers and flares. The '76—'78 Mustang-II Cobra is an example that didn't work. It had the graphics and body trim, but was sorely lacking in the handling and go departments. Even with a V8 engine, it remained a Pinto at heart. Chrysler tried to revive the Road Runner based on its F-Body Volare, but the old magic was gone.

Chevy even tried a junior Z/28 based on the Vega-derived Monza. The Monza Spyder was available with a small-block V8, sport suspension, plus all the usual decals, scoops, spoilers and fender flares. It was a dismal flop!

The Trans-Am had just the right blend of performance, handling, styling and *continuity of image*. It had been around long enough to build up a following. Pontiac sales people never downgraded that factor. Forging an image takes time, and a lot of right guesses.

Pontiac and John Schinella *did* find a substitute for brute performance in the '70s.

Chevy went the flare-and-graphics route with the Monza Spyder of the late-'70s. Sales were disappointing. Photo courtesy of Chevrolet.

1976 Mustang Cobra II model was typical of the "spoiler-and-flare" period in the mid-'70s, when the manufacturers supplied the youth market with radical looks rather than performance. The cars looked faster than they actually were. Photo courtesy of Ford Motor Co.

Ford tried the wild hood-graphics routine with the 255-CID '80 Mustang Cobra. Photo by Tom Monroe.

Chevy Monza had one good thing going for it—excellent aerodynamics. The HPBooks Monza, with an injected 370-CID small-block went over 219 mph at Bonneville in 1981. Car holds four records.

Long-awaited third-generation Camaro debuted in '82. Together with the Firebird and Mustang/Capri, it led a performance revival in Detroit. Photo courtesy of Chevrolet.

Supercars in the '80s

Detroit's Performance Renaissance

THE SITUATION IN THE '80s

Engineering and marketing conditions in the American auto industry in the early '80s were vastly different than those of preceding decades. For one thing, exhaust-emission restrictions tightened down even more—to the point where a simple catalytic converter no longer could do the job. This necessitated some type of feedback system to sense the completeness of combustion in the exhaust gas ahead of the converter. With most systems, the signal went back through an electronic computer circuit to adjust the air/fuel ratio at the carburetor or fuel-injection unit. This added tremendously to the cost and complexity of the engine.

Another factor was the continually rising cost of gasoline. This, combined with federal Corporate Average Fuel Economy (CAFE) standards and the specter of a gas-guzzler tax levied on certain performance models, forced designers to re-evaluate fuel efficiency in all price and size ranges. These considerations forced the trends to smaller-displacement four- and six-cylinder engines, tires with less rolling resistance, four- and five-speed transmissions, lockup-converter automatic transmissions, lightweight materials, better body streamlining and so on.

Still another important marketing factor in the '80s was the increasing sales competition from imported cars. Imported cars had been taking a bigger chunk of the U.S. market for years. Most of the competition in the '70s was in the area of small economy cars.

But at the beginning of the '80s, the Japanese especially began to penetrate the remnants of the sporty/performance-car market. The Datsun and Mazda offerings, and more recently the sporty models from Toyota and Mitsubishi have been pretty attractive—in styling, price *and* performance. And keep in mind that *performance* here means cornering, stability and braking as well as speed and acceleration.

After what seemed an eternity, Detroit responded with sporty/performance cars once again. This time, though, it was a curious mix of traditional V8-powered rear-drive cars, four- and six-cylinder front-wheel-drive cars, and turbocharged cars. Never mind that none of these were capable of 14-second quarter-mile times. These cars were and are fun to drive!

Perhaps the most-convincing examples of Detroit neo-supercar sports coupes were the Mustang GT/Capri RS, and Z/28 Camaro/Firebird Trans-Am. These pairs of cars shared the same basic body and chassis, with major differences in trim, fittings and suspension details. All were designed to give outstanding cornering, handling, stability and braking. All of this was done in a styling envelope that combined low aerodynamic drag with the long-hood, short-deck silhouette trademark of the Corvette and the first ponycar—the '64-1/2 Mustang.

Above: Cutaway of the 1982 F-body Firebird. Bucking a trend to front-wheel drive, the cars were front-engine, rear drive. Emphasis was on handling and fun driving rather than brute acceleration. In Trans-Am trim, car was capable of 0.82 G lateral acceleration. Photo courtesy of Pontiac.

For '83, Chevrolet offered a five-speed overdrive transmission in the Camaro. Photo courtesy of Chevrolet.

1982 Firebird at speed shows clean lines that gave aerodynamic drag coefficient of 0.32. Photo courtesy of Pontiac.

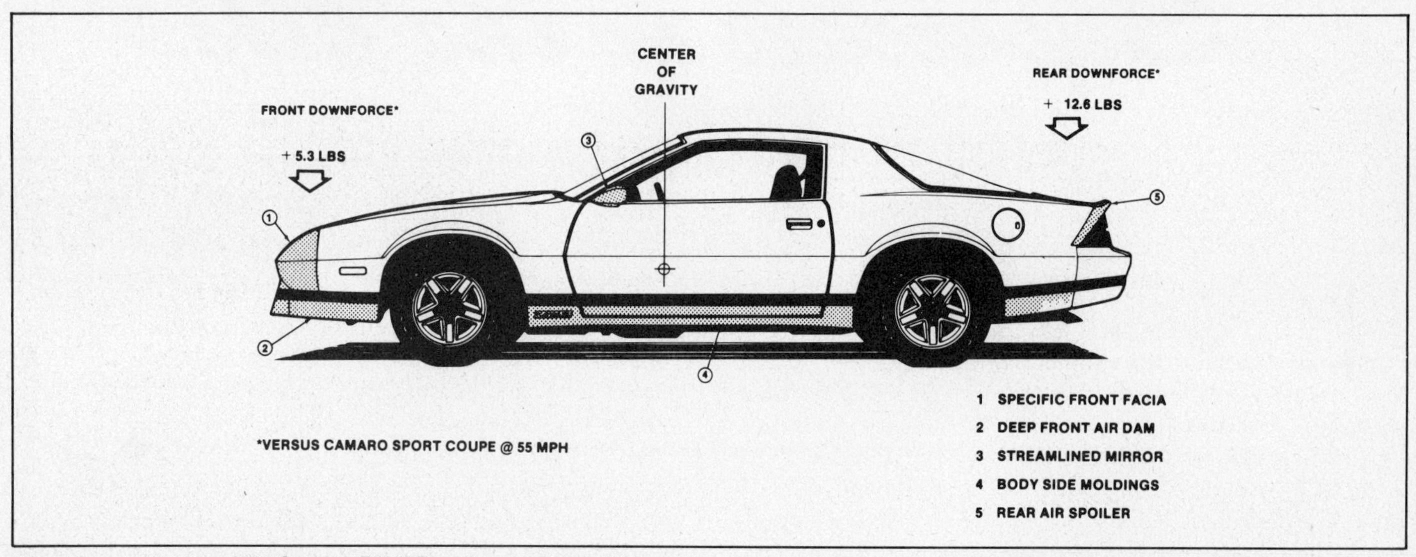

Aerodynamic details of '82 Camaro Z/28. Drawing courtesy of Chevrolet.

But note carefully: Straight-line speed and acceleration were secondary considerations all the way. Detroit's new sports/performance cars were designed for a new performance theme—more-responsive handling in a trim, lithe machine with reasonable fuel efficiency.

Camaro Z/28 and Firebird Trans-Am—This pair of F-cars was introduced in '82. The cars were indicative of GM's new thinking. For one thing, front-wheel drive was rejected because this overworks the front tires during cornering. And, on a performance car, rear-wheel drive is desirable because the driver can get oversteer on demand by blipping the throttle. Dirt-track racers have used this technique for years.

Also, an independent rear suspension was rejected because the potential handling improvement didn't justify the cost and extra weight. The reduced unsprung weight of IRS helps mostly on rough roads—not a major consideration on a medium-priced street car. A *live* rear axle can give excellent handling with the right suspension geometry and tuning. And by moving the engine back a little, GM was able to achieve a near 50/50 weight distribution. Weight distribution is the best way to achieve balanced cornering with the same size tires.

Coil springs were used at all four corners. But suspension-linkage layouts were unusual at both ends. At the front, the coil spring acted on the lower control arm, but vertical wheel motion was controlled by a telescoping strut containing the shock absorber. It was an attempt to combine the best features of the classic unequal-length-control-arm suspension with the space- and weight-saving MacPherson-strut layout. This setup followed the earlier design used on Fairmonts, Mustangs, Thunderbirds and Granadas.

At the rear, the axle was carried on short pivoting trailing arms, with a Panhard rod for lateral control. But then there was a 5-ft-long torque arm extending from the center of the axle housing—to which it was solidly bolted—to the rear of the transmission. Its effect was to resist axle twist

Chevrolet's unique TBI *Crossfire* V8 of the early '80s used electronic throttle-body fuel injection to maximum performance with minimum emissions. Photo courtesy of Chevrolet.

Factory engineers got some performance kicks in the '80s by furnishing special high-speed models to pace the Indianapolis race. This is a 250-HP Camaro Special. Photo courtesy of Chevrolet.

Ford rekindled muscle-car fever in '82 by dropping a high-output 302-CID V8 into the Mustang. The resulting car was the quickest Detroit machine in the early '80s. This is the 177-HP '83 Mustang GT. Photo courtesy of Ford Motor Co.

on acceleration and braking. This design was similar to an old *torque-tube drive,* but without the weight and complexity. The torque arm was a feature of the Chevy Vega and subsequent Monza-based cars. The torque arm controlled axle windup and eliminated the need for a rear-axle upper control arm. It also allowed more space for the fuel tank and luggage compartment.

The GM F-cars could be ordered with P215/65R-15 Goodyear Eagle GT radials with 7-in.-wide treads mounted on 15 X 7 in. aluminum wheels. In conjunction with higher-rate springs, stiffer shocks and anti-roll bars, these tires raised maximum cornering capability to as high as 0.82 G. The highest cornering figure recorded on the Trans-Am ponycars of 1970 was 0.73 G! When you consider that the finest and most-expensive imported sports cars rarely exceed about 0.85 G, the achievement of Camaro chief engineer Tom Zimmer and his counterparts at Pontiac really stand out.

Braking was competitive too. An optional brake system used 10.5-in. vented discs at all four wheels, designed for

Suprisingly quick 5-liter '83 Capri RS shared top-acceleration honors with its Ford sibling. Photo courtesy of Ford Motor Co.

A sign of the times. By the late '70s and early '80s, standard police-car performance was so bad that law-enforcement officers couldn't overtake and intercept most of the earlier-model cars on the road. The fix was for the highway patrol to buy muscle cars themselves! This '82 5-liter Mustang GT four-speed is decked out in full Arizona Highway Patrol gear. Photo by Tom Monroe.

fade-free stopping in the most-spirited road driving. Vacuum boost kept pedal pressures under 60 lb. With the wide-tread tires for traction, and the long torque arm to prevent wheelhop, stopping distances from 70 mph were among the best in the world at about 201 ft.

GM went for maximum performance with minimum emissions and fuel consumption by using an electronically controlled fuel-injection system. The system metered fuel spray from a *throttle-body* injector unit mounted on a conventional cross-ram intake manifold. Two injector units were used on the manifold to get an extra boost in midrange. Total output was 165 HP—up from the 4-barrel carbureted version's 145 HP.

Fuel metering was by a feedback network that constantly monitored oxygen content in the exhaust to select the most-efficient air/fuel ratio. And the electronic control box that measured the fuel was also used to monitor detonation, control transmission shifts, provide various function readouts on the instrument panel—and a dozen other jobs.

Mustang GT/Capri RS—GM's new-for-'82 F-cars created a stir in the youth/performance market, but Ford wasn't going to let them steal the show. Ford had a lightweight, highly developed contender waiting in the wings. All it needed was more engine. With Ford's Escort and Lynx econoboxes selling in big enough numbers to satisfy federal corporate fuel-economy standards, the boys from Dearborn had a chance to do something they hadn't been able to do in nearly a decade—hop-up a V8.

Ford engineers applied some hot-rod science to their basic 5-liter V8. A little-more carb, a little-more cam, better-flowing exhaust. With existing off-the-shelf performance equipment, such as a *marine* camshaft that had been relegated to "off-road" use only, they got 157 HP for '82. The following year, they complemented the package with a 600-cfm Holley 4-barrel carb on an aluminum manifold. The move from the Motorcraft 2-barrel to the Holley helped bump output to 177 HP.

When Ford finished the transformation, the cars were 200—400-lb lighter than the GM F-cars. Result: The Mustang GT and Capri RS models could run circles around the

Car	Curb weight	Trans	Final Drive	Engine	Adv. HP	0–60	1/4-mile e.t.	Braking 70-0	Skid-pad
Z/28	3440	A3	3.23	305 TBI	165	8.6	16.4 @ 83	201	0.81
Mustang GT	3000	M4	3.08	302 2-bbl	157	8.1	16.2 @ 86	213	0.75

Z/28 and Trans-Am on acceleration—with figures even approaching supercar times of the '60s! Check out the above *CAR and DRIVER* road-test figures for the '82 models:

For the Mustang GT and Capri RS, Ford recalibrated the suspension for more-neutral handling characteristics. And with the TRX suspension included with the GT and RS packages, and matching 190/65 HR-390 Michelin tires, handling was competitive, too. Although front-heavy, the 55.8/44.2 front/rear weight bias was better than any previous Mustangs of the supercar era. These were definitely the hottest-performing models available from the American auto industry in the early '80s.

NEW JUNIOR SUPERCARS

When I last talked about junior supercars, they were '60s-vintage compact and intermediate cars with smaller engines and fewer frills. And with the possible exception of the Plymouth Road Runner, they were not successful sellers. By and large, these cars were rear drive with V8 engines. Sure, there were the turbocharged Corvair and the OHC-Six Tempest Sprint, but these were short-lived exceptions.

In the late '70s and early'80s, Detroit began reintroducing these types of cars. But this time, it was with an eye towards maintaining good fuel economy. So this crop of junior supercars had four- and six-cylinder engines. And because some were based on econoboxes, many of the cars had front-wheel drive.

As in the '60s, there were many junior-supercar pretenders with fake scoops, stripes and spoilers. But here are a few I think worth mentioning:

Dodge Charger 2.2—With Chrysler Corp. teetering on the verge of bankruptcy in the late '70s, they had to make a number of promises to Washington in order to qualify for federal loan guarantees. In the atmosphere of the downfall of the Shah of Iran in '79 and the resulting surge in oil prices and import-car sales, Chrysler promised to move to fuel-efficient, small, front-wheel-drive cars. This made their proven performance machinery of the previous decades obsolete.

So Chrysler spun a sports coupe model off their econobox Omni/Horizon. The resulting Plymouth Horizon TC3 and Dodge Omni 024 cars offered a good mix of sporty styling, decent handling with the optional suspension package and satisfactory performance. In '79—'80, the only engine available was the 80-HP 1.7-liter OHC four—supplied by Volkswagen. But in '81, the new Chrysler-built 2.2 OHC four became available. It woke up performance.

To give these cars the image and recognition they needed to enter the youth/performance market, Chrysler tapped—you guessed it—Jim Wangers. With Wangers

With Chrysler committed to front-wheel drive and four-cylinder engines, pundits declared the MoPar muscle car dead. Wrong! The Dodge Charger 2.2 and Plymouth Turismo 2.2 proved to be the junior-supercar pocket rockets of the early '80s. Photos courtesy of Chrysler Corp.

After enjoying a modicum of success with the Charger and Turismo, Chrysler decided to enhance their performance and image. In connection with Carroll Shelby, Chrysler developed the Dodge Shelby Charger. Photo courtesy of Chrysler Corp.

help, Chrysler combined the engine and suspension options with the usual scoops, spoilers, stripes and flares. The resulting Dodge Charger 2.2 and Plymouth Turismo models proved to be sales successes. The little four-banger, breathing through a low-restriction air cleaner and exhaust system and driving through a numerically high 3.56:1 final-drive could move these 2360-lb cars in fairly brisk fashion—especially out of the hole. Acceleration—0—50 in 6.6 seconds—was respectable for a four-cylinder.

In 1983, Chrysler recontoured the cylinder-head ports and manifold passages and netted a 10-HP gain—bumping output to about 100 HP. The result was 0—60-mph times in the 8-second bracket.

But with the Mustang GT and Camaro Z/28 heating up the performance market, Chrysler couldn't rest on its laurels. Lee Iacocca coaxed none other than Carroll Shelby away from his chili-bean business to cook up more performance goodies for the Charger. Iacocca hoped Shelby's performance reputation would rub off on his little MoPar.

The resulting '83-1/2 Dodge Shelby Chargers were exciting. Engine output was boosted to 107 HP with a higher 9.6:1 compression ratio, high-lift cam, modified spark-advance curve and a unique air-pump cutout at WOT. Final-drive ratio was raised to 3.87:1. Higher-capacity brakes from the heavier K-car were used with vented front rotors. High-rate suspension springs were used. Outside, the car sported an air dam with ground-effects skirts. The Shelby Charger was a visually exciting, good-handling car, with suprising performance and good fuel economy.

Another front-wheel-drive junior supercar that established a reputation for itself was the GM X-car. This is a Chevrolet Citation X-11. Photo courtesy of Chevrolet.

GM X-Car Sport Coupes—Chevrolet took the first step toward a high-performance front-drive compact with the 1980 Citation X-11 model. It was such a success that the other GM divisions came in with the Buick Skylark T Type, Pontiac Phoenix SJ, and the Olds Omega SX. All of these used the *HO*—high output—version of the 2.8-liter, 60° Chevy V6. The V6 got 135 HP at 5400 rpm with oversize valves, higher compression, high-lift cam and low-restriction exhaust. This was enough for 0—60-mph times in the 9-second range.

But the precise handling of these cars was perhaps even

When fitted with the high-output version of Chevrolet's 60° 171-CID (2.8 liter) V6 and sport suspension, the Phoenix SJ was an impressive performer. Photo courtesy of Pontiac.

Buick didn't want to get left out in the cold without a high-performance X-car of its own. This is the '83 Skylark T Type. Photo courtesy of Buick.

more exciting than their acceleration—especially considering the front-wheel drive. The GM designers matched their suspensions to the remarkable response and cornering power of the Goodyear Eagle GT radial tires, to achieve excellent handling. The result was a neutral feel right up to breakaway, and a pleasing car to drive. With the 6000-rpm capability of the engine, a good four-speed manual transmission—cable-shifter notwithstanding—and the tremendous grip of the Eagle GT tires, there was as much driving fun in this car as you could find anywhere.

Mustang SVO—The SVO introduced in mid-1983 was an altogether different car than the disastrous Mustang Turbo of 1979. This time, Ford took great pains to develop the 2.3 four-cylinder into a *performance* turbo engine.

Why bother with the 2.3? Whereas performance could be more-easily squeezed from the V8, the four-cylinder promised to deliver a better performance/economy compromise. In addition to turbocharging, specific output was boosted with direct-port fuel injection. An intercooler was used to cool the compressed charge—like all proper turbos! Acceleration was not as strong as the V8s, but overall fuel economy was somewhat better.

In accordance with Ford's emphasis on aerodynamics, the SVO Mustang also featured more-extensive use of spoilers to achieve lower drag as well as wild styling lines.

GRAND-TOURING CARS

I'm introducing this classification for the first time. Long a European tradition, grand-touring cars blend four-passenger comfort, high-speed cruising cabability and superior response and handling into a sleek package. In the late '70s and early '80s, Detroit began to produce this type of car.

Buick Regal Turbo—When Buick resurrected its '60's-vintage V6 in 1975, it was for one reason—to boost fuel economy in a package that would appeal to traditional Buick-V8 buyers. But the Buick ad people, McCann-Erickson, found a way to sell the V6 as the perfect blend of performance and economy.

Those first 225- and 231-CID V6s had miserable idle

Buick made a silk purse out of a sow's ear by turbocharging their 231-CID (3.8 liter) 90° V6 in '78. The resulting Buick Regal Turbos were responsive road machines. This is an '83 model. Photo courtesy of Buick.

Buick's 231-CID V6 Turbo used an AiResearch turbocharger with 5—7-psi boost. A detonation sensor backed off spark advance at the onset of detonation to protect the engine. Photo courtesy of Buick.

Ford stood Detroit watchers on their ear with the reskinned '83 Thunderbird. A complete break from the previous Bordinat styling tradition, the new 'Bird had distinctly European lines with an eye towards aerodynamics. Shortly after its introduction, Ford made available the 5-liter HO V8 and 2.3 SVO Turbo Four engines. Photo courtesy of Ford Motor Co.

quality—thanks to the uneven firing pulses of the 90° configuration. But in mid-'77, they *splayed*—offset—the crankpins to get even firing pulses. Committed to the V6 as their volume engine of the '80s, Buick began doing some serious performance development work on it.

In '78, they introduced their first turbocharged versions in the Century and Regal—one with a Rochester Dualjet and the other with a Quadrajet. Fuel economy was no better with the deuce, so it was dropped. The 4-barrel turbo Buick V6 has been with us ever since.

Of all the GM A-Special bodies of the late '70s and early '80s, the Buick Regal proved to have the slickest aerodynamics and was therefore the most popular among GM NASCAR racers. Photo courtesy of Buick.

Early '80s were lean years for Ford lovers on the NASCAR tracks. But the slick aerodynamics of the '83 'Bird and the re-release of 351 Cleveland engine parts promised a challenge to the GM-dominated high-banked ovals. Photo courtesy of Ford Motor Co.

An interesting feature was the way Buick engineers attacked the detonation problem. With the low-octane fuel available in the '70s, turbo boost led to detonation. This would destroy the engine in short order. Buick used a knock sensor that automatically retarded spark timing whenever detonation occurred. The system worked, and Buick V6 turbos have proved to be durable.

Continuing development under Buick General Manager Lloyd Reuss made it into a sophisticated performance package with good fuel economy. The late 3.8-liter (231-CID) version developed 180 HP, which was good for 0–60-mph times in the 8-second bracket with the luxury Regal coupe. For 1983, the turbo engine was combined with the "Euro Spec" T Type suspension, to give sports-car handling and response in a plush personal car. It was not a true high-performance car even in today's terms. But Buick demonstrated that turbocharging could be tamed to give power with almost any degree of smoothness and luxury.

Thunderbird—No doubt about it, Ford set many a Detroit watcher on his ear with their breathtaking '83 Thunderbird. A dramatic reversal from previous T-Bird designs, the '83 model was a showcase of aerodynamic engineering.

It was once thought that reducing wind drag on a body would compromise styling. Ford proved this to be untrue with the '83 T-Bird. The wind tunnel C_d *factor*—drag coefficient—was only 0.35, with the lines smooth and flowing from front to back. This refinement boosted fuel economy about 1–2 mpg at 60-mph highway speeds. And with a high-output 5-liter V8 or Turbo Four under the hood—the top speed was 10- or 15-mph faster than many of the old '60s supercars with that same amount of horsepower. Performance is where you find it!

In fact, the shape was so promising that NASCAR racers flocked to the new design in '83. If the shape helped fuel economy at 50–60 mph, imagine what it would do for road horsepower at 100 mph!

Of course body shape wasn't the whole story. Ford management woke up and took advantage of existing tooling—even though it was in Australia—on the 351C four-bolt block and big-port heads. The 351C engine in the slippery '83 body proved to be a potent combination on the high-bank ovals of NASCAR.

Hurst Olds—Back in 1968, Oldsmobile got around the GM corporate ban on 455-cube engines in intermediate bodies by getting the Hurst specialty people to do it for them. Olds dealers sold the car as a special limited-production model. Those early Hurst/Olds models are valuable collector cars today.

With performance back in vogue for '83, Hurst decided to do it again. Again it was a Cutlass coupe, and again that catchy black-and-silver paint scheme. Not so much engine this time—only the 307-CID version of the Olds V8 was available. But the 5-liter V8 was hopped up to 180 HP with a marine cam, 4-barrel carburetion and low-restriction dual exhaust. This was enough for 0–60-mph times of 8.5 seconds with the 3500-lb car.

Handling was impressive with Goodyear Eagle GT tires and tuned suspension. One of the most-dazzling gimmicks in the driver environment was a unique Hurst console shifter with a separate lever for each manual upshift with the four-speed automatic. You could drag through the gears, selecting your own shift points, or rack all the levers for automatic shifting. A trendy takeoff on Hurst's new Lightning Rods competition shifter.

THE AMERICAN SPORTS CAR

Corvette—In the Spring of '83, Chevrolet finally introduced its long-awaited "new" Corvette—the first all-new Corvette in 15 years! The car proved worth the wait. Zora Arkus-Duntov would've been proud. It was a car virtually without compromise!

The "old" 'Vette was 1968-vintage technology—that is to say it was heavy, cramped, and IRS notwithstanding,

After a production run of 15 years, the third-generation Corvette was laid quietly to rest. The new '83 Corvette proved worthy of the title, "The American Sports Car." Photo courtesy of Chevrolet.

Corvette's clamshell, flip-up hood gave good access to engine components. Available was a five-speed manual transmission with overdrive in the top three gears. Photo courtesy of Chevrolet.

1983 Corvette chassis included a "backbone" drive train "C"-section beam. Lightweight aluminum was used for major suspension components, with fiberglass transverse leaf springs front and rear. Photo courtesy of Chevrolet.

didn't handle as well as many live-rear-axle cars—such as its Z/28 stablemate. In the late '70s and early '80s, the 'Vette was the subject of many jokes—even within GM engineering ranks. With a curb weight approaching two tons, it was often described as the "world's fastest truck."

Introduced in the Spring of '83, the *1984* Corvette shed nearly 300 pounds—something the 205-HP throttle-body-injected 350-CID small-block Chevy V8 appreciated. How do 0—60-mph times in the 6-second bracket grab you? Here was a car that could accelerate as well as the High-Output 5-liter Mustangs and Capris, using basically the same induction system as the Z/28.

Where the new Corvette excelled was in handling. An optional handling package used a radical new P255/50VR16 Goodyear *unidirectional* tire on lightweight cast-aluminum 16-in. wheels, in conjunction with tuned suspension and extra roll stiffness. The front wheels were 8-1/2-in. wide, and the rears an incredible 9-1/2-in. wide. Early skidpad testing indicated remarkable roadholding—over 0.9 G!

WHERE TO FROM HERE

Most car enthusiasts agree that the days of Detroit's emphasis on brute, straight-line performance—as experienced in the '60s and early '70s—are gone forever. The high price of gasoline alone has seen to that—not to mention the restrictive effects of exhaust-emission regulations. And as far as that goes, federal "safety Nazis" seeking to protect everyone from every danger known to man seem intent on stopping any substantial improvement in car performance.

But all of this doesn't mean an end to fun driving. Cars with a balanced mixture of responsive handling, braking, power and interior comfort can be fun to drive. Sure, the old muscle cars had their heavy-duty suspensions that gave better than average handling—on smooth roads. Today, Detroit is learning to tune the whole chassis to work as a system—from the rigidity of the body structure to the coupling between roll-bar rates and tire slip-angles.

Suspension engineering is a whole new science today. And the resulting cars are a new experience to drive. They're quick, agile, stable and *fun.* "Responsive" is the watchword.

This doesn't imply that there won't be future improvements in performance. There are many ways acceleration can be improved within the framework of high fuel prices and tight emission restrictions: engine turbocharging, electronic fuel- and spark-control systems, variable-ratio transmissions, lighter materials, slick aerodynamics and so on. Progress will come. But all these things involve greater *complexity* and *cost.*

Whether Detroit's offerings ever match supercar acceleration again is not important. What's important is that the manufacturers give us cars that are *fun.* With fun cars once again emanating from the Motor City, there will be ample enthusiasm for another *American Supercar* era!

AAA—American Automobile Association
All-out—1. Full-scale competition car 2. Maximum car speed
Altered—Competition class permitting extensive modifications to body and engine
At-speed—Operation of a car at top speed
Bad scene—No good
Banzai—All-out run, bringing car to peak of performance
Bash—Race or event
Beast—1. High-performance hot rod 2. Difficult car to drive
Big bore—Engine with larger-than-normal cylinder diameter
Big end—1. End of quarter-mile 2. Crankshaft end of connecting rod
Bite—Tire traction
Blow—Engine failure
Blow-off—To pass a car decisively when racing
Blower—Supercharger
Blown—Supercharged
Board speedway—Race track with wood surface
Boattail—Body style popular in '20s and '30s with tapered, boat-like rear-end treatment
Bogie—Goal or target
Bore out—To increase engine displacement by increasing cylinder bore
Bored & stroked—Cylinder bore and crankshaft stroke increased
Bored—Cylinder bore increased
Boss—1. Ultimate 2. Top quality
Bottom-end—Cylinder-block main-bearing webs, crankshaft and rods
Boundary layer—Layer of slow-moving air adjacent to surface of moving body
Box—Gearbox; usually used in reference to manual transmission
Bread and butter—Stock
Bungee cord—Elastic cord with hooks on ends used to tie down equipment
Cc—Cubic centimeter; metric measurement used for engine displacement (100cc = 1 liter = 61 cubic inches) (16.38cc = 1 cubic inch)
C_d—Coefficient of drag
Cam—Camshaft
Cammer—Overhead-cam engine
Cheaters—Slick-tread rear tires used in competition events
Clean—Well-built car; sanitary
Clock—1. Slang for speedometer or tachometer 2. To time a car's performance
Cobbled—Temporary, and usually crude, modification
Come unglued—Part or assembly breaks or comes apart
Corner—Curve on a race track
Cowl Induction—Cold-air engine inlet drawing air from high-pressure area at base of windshield
Crank train—Crankshaft, pistons and rods
Crank—Crankshaft
Crew chief—Mechanic in charge of pit operations
Cross ram—Ram-induction system, usually with diagonally-opposite carburetors and longer-than-stock runners; used on Chrysler 426 race Hemis and some Camaro Z/28s
Cubes—Engine displacement in cubic inches
Cutout—Covered opening in exhaust system ahead of mufflers, easily removed for racing
Deuce—1. 2-barrel carburetor 2. 1932 Ford
Dig out—To accelerate rapidly from a standing start
Doctor—To modify
Drafting—Following closely behind another car to lower aerodynamic drag: for conserving fuel while maintaining speed or increasing speed for passing
Drag strip—1. Quarter-mile race course with deceleration area, usually specified as 60-ft wide and 4000-ft long 2. Any paved area used for straight-line acceleration contests
Drag—Two cars engaged in a standing-start race
Dragster—1. Car built exclusively for drag racing 2. Driver who participates in drag racing
Driveability—Desirable all-around performance characteristic; engine is tractable when cold and at low speeds
Dual quad—Carburetor setup using two 4-barrel carburetors
Duesie—Duesenberg
Dumps—See *cutouts*
ET—Elasped time used in drag racing, road races, rallies
Eliminated—Beaten in a drag race
Eliminator—Drag car that wins by eliminating other cars in its class
Export parts—Manufacturer's name for *racing parts;* also called *off-road, police,* or *severe-usage parts*
Eyeball—To look over something
F-head—Engine configuration with at least one side valve and one overhead valve per cylinder
FI—Fuel injection; fuel-feed system system where fuel is sprayed directly into engine ports, combustion chamber or throttle body rather than drawn through a carburetor
FX—Factory Experimental class
Fastback—Car with sloping back window
Fat top end—Good power at high rpm
Flat out—Driving at top speed
Flat spot—Point at which increased throttle opening momentarily fails to increase engine rpm
Flathead—Engine with valves in the block; L-head, T-head, or side-valve engine
Flog—To drive a car badly or hard
Flying start—In racing, a start made at speed
Four-banger—Four-cylinder engine
Four throat—4-barrel carburetor
Four-bbl (4-bbl)—4-barrel, or 4-venturi carburetor
Four-on-the floor—Four-speed floor shift
Four-speed—Four-speed manual transmission
Full bore—Throttle wide open, flat out
Full house—Car (or engine) with all possible performance modifications short of supercharging. Also called *full race*
G.T.—Gran turismo; high-performance production car that accommodates two people and luggage, and is capable of over-the-road touring or class racing
Get out of it—To lift foot off accelerator
Goat—Pontiac GTO
Goodies—Hot-rod accessories, engine modifications, rare or valuable parts
Grand National—NASCAR late-model stock-car circuit for intermediate and full-size cars
Groove—Fastest part of a race track
Guzzler—Car with high fuel consumption
Hairy—1. Potent-performing car 2. Difficult race course 3. Poor-handling car
Handling—Car's chassis performance
Hang out the rear—To negotiate a corner or curve with the rear wheels in a controlled skid
Hauler—Fast car
Header—Racing-type exhaust manifold
Hemi—Engine with *hemispherical* combustion-chamber cylinder heads
HiPo—High-performance 289-CID Ford
Hopped-up job—See *Special*
Hot rod—Passenger car modified by individual with aftermarket or other non-stock parts to improve its performance
Hydraulic cam—Camshaft used with hydraulic valve lifters
IRS—Independent rear suspension
Idiot light—Derogatory term for engine warning light
Index of performance—Evaluation system in racing that mathematically considers engine size, car weight, efficiency and finishing position: thus, a car may win *on index* without being the fastest car
Injector—Fuel injector
Iron—Conventional car (as opposed to sports or high-performance cars)
Jug—Carburetor or cylinder
Junior supercar—Low-priced supercar, usually with smaller engine and lower weight
Kettering engine—Olds/Cadillac OHV V8 engine introduced in 1949
Knock-off—Wheel-lug nut that can be quickly removed
L-head—See *flathead*
Lay rubber—To spin tires on pavement
Light the tires—To spin tires so they smoke
Liter (litre)—Metric measure of engine displacement (1 liter = 1000cc = 61 cubic inches)
Locker rear end—Rear axle that prevents one wheel from spinning independent of the other
Long ram—Chrysler ram-induction system with 30-in.-long passages
Loud pedal—Accelerator

Low end—1. Low-rpm torque 2. Low-rpm performance characteristics
Lunch—To destroy an engine
Mag—Magneto
Mags—Magnesium wheels
Mains—Main bearings
Mill—1. Engine 2. To remove metal from the base of cylinder head or block-deck surface
MoPar—Hot Chrysler product; derived from Chrysler *Motor Parts*
Model B—First Ford V8 of 1932
Motor City—Detroit
NASCAR—National Association for Stock Car Auto Racing; world's largest stock-car race-sanctioning body
Neutral—Balanced handling, as opposed to understeer or oversteer
Off the line—Start of race
Open exhaust—No mufflers
Out of the gate—Start of drag race
Out of the hole—Start of drag race
Over-rev—To rev an engine beyond its rpm limit or redline
Overbore—Engine with cylinders enlarged over stock diameters
Oversquare—Engine with bore greater than stroke
Oversteer—The tendency of a car to lose rear-wheel traction while cornering at speed
Pace car—Car that leads pack for one or two laps preceding race starts; also paces race under caution flag
Parts-bin special—Car assembled from existing production parts
Peak out—To rev engine to its limit
Peaky engine—Engine that develops good power only within narrow rpm range
Peel, peel rubber—To spin tires so rubber is deposited on road surface
Pipes—Exhaust pipes
Pit crew—Team that services a race car during a race
Pits—Trackside area for servicing race cars
Plunge-cut—Machining operation to flathead engine block to unshroud valves for improved breathing
Police equipment—See *Export parts*
Poncho—Hot Pontiac
Ponycar—Sporty compact car with long-hood, short-deck styling and 2+2 seating
Pop the clutch—Sudden clutch engagement
Porcupine-valve—See *Semi-hemi*
Port—1. Openings in a block or cylinder head through which air/fuel mixture enters or exhaust leaves the combustion chamber 2. To modify, usually by enlarging valve passages for improved engine breathing
Power (wheel) hop—Tendency of rear wheels to shudder or hop under maximum acceleration
Power pack—simple engine, drive-line and, sometimes, chassis package offered by manufacturer to improve performance
Progressive linkage—Linkage for multi-venturi and multiple-carburetion systems that allows staged opening of venturis, or carburetors: Designed to provide greater air/fuel flow capacity as engine rpm increases
Q-Jet—Rochester Quadrajet carburetor
Quad—4-barrel carburetor
Ragtop—Convertible
Rake—(California)—Car with front end lower than rear end
Rake—(Cowboy)—Car with rear end lower than front end
Ram air—Induction system with carburetor air intake in high-pressure area, usually at front of car
Ram induction—Intake manifold system with runners tuned to utilize oscillating pressure waves and supercharge intake charge at specific rpm range
Rat—1. Bad-running car 2. Big-block Chevrolet engine
Redline—Engine-rpm limit
Relieved—Passages cut into cylinder block to increase airflow of a flathead engine
Revs—Revolutions per minute, or rpm
Roll bar—Steel tubing that protects driver in case of rollover
Roller—Camshaft setup with roller tappets
SCCA—Sports Car Club of America
Sanitary—1. Unusually-clean car 2. Well-prepared race car
Scoop—Opening in body to deliver outside air to engine, brakes or cockpit
Screamer—1. High-rpm engine 2. Fast car 3. Supercharged engine
Semi-hemi—Canted-valve layout as first used in big-block Chevy
Set up—To prepare or modify a car for racing
Shaker—Air scoop attached directly to carburetor air cleaner and protruding through gasketed opening in hood
Short ram—Chrysler ram-induction system with 15-in.-long passages
Six-Pack—Chrysler product with three 2-barrel carburetors
Six-banger—Six-cylinder engine
Skins—Tires
Sled—Car
Sleeper—1. Racing car that performs better than expected 2. Stock-appearing car with better-than-stock performance
Slick—Treadless drag-racing tire, usually with wide cross section
Solids—Mechanical valve lifters
Soup-up—To change or modify an engine to increase power
Special—1920's and early '30's term for hot rod
Speed shift—To shift gears rapidly without lifting off the accelerator
Speed shop—Store that sells high-performance parts to public
Speedster—Two-seat open-bodied roadster with low-slung frame rails
Standing quarter—Quarter-mile timed race begun with vehicle at rest
Stick shift—Manual-shift transmission
Stocker—1. Stock car 2. Stock-car driver
Stovebolt—Six-cylinder Chevrolet
Strippo—Car with all frills removed
Stroke—Distance crankshaft causes each piston to travel in its cylinder
Stroker kit—Crankshaft and connecting-rod assembly engineered to increase engine displacement by increasing piston travel
Stroker—1. Driver who keeps steady pace 2. Longer-stroke crankshaft
Strong—Powerful engine
Super stock—Production car with engine and chassis modifications to run in specific drag-racing class
Supercar—Passenger car modified by manufacturer for above-average performance
T-head—Flathead
Tach—Tachometer
Throat—Carburetor opening at throttle valve
Tip-in response—Initial throttle response
Top eliminator—Only car remaining in a class after elimination runs
Top end—Power output at high rpm or at end of quarter-mile
Top time—Lowest quarter-mile e.t.
Torque-up start—Method of launching automatic-transmission-equipped cars in drag racing: Driver applies brake with transmission in forward range and depresses accelerator. Brake is released to start
Traps—Three-light system at the end of the quarter-mile that stops the elapsed-time clock for recording e.t. and top speed
Tri-Power—Pontiac name for triple 2-barrel carburetion
Tri-Y—Exhaust-header configuration with 4 into 2 into 1 primary pipes
Trick—New, unconventional modification
Trips—Triple 2-barrel carburetion
Tweak—To modify, hop up, soup up
USAC—United States Auto Club; sanctioning body for Indy Car races as well as sprints, midgets and stock cars
Undersquare—Engine with stroke greater than bore
Understeer—Condition where the front tires of a car tend to lose traction before the rear tires while cornering at speed
Unreal—Exceptional, outstanding
Valve float—Engine rpm at which valve springs cannot shut the valves
Velocity stack—Tube extending up from air horn of a carburetor or fuel injector to increase airflow
Ventilated block—Engine block with hole caused by engine failure
Venturi—Narrow section of carburetor throat that increases air velocity to draw in fuel from float bowl
Wail—To run fast
Wedge—1. Raise or lower one corner of a race car to shift weight and improve handling 2. Combustion-chamber shape
Wild—Car that deviates greatly from stock appearance and/or performance
Windage tray—Baffle which prevents crankshaft from whipping oil

Index

4.7162078368